Science and
the Cure of Diseases

*Letters to Members
of Congress*

Efraim Racker

Science and
the Cure of Diseases

*Letters to Members
of Congress*

*Princeton University Press
Princeton, New Jersey*

Copyright © 1979 by Princeton University Press

Published by Princeton University Press, Princeton, New Jersey
In the United Kingdom: Princeton University Press,
Guildford, Surrey

All Rights Reserved

Library of Congress Cataloging in Publication Data will be
found on the last printed page of this book

This book has been composed in VIP Century Schoolbook

Clothbound editions of Princeton University Press books
are printed on acid-free paper, and binding materials are
chosen for strength and durability

Printed in the United States of America by Princeton
University Press, Princeton, New Jersey

Dedicated to
Franziska and Ann,
my medical advisers

Contents

Preface

Scientists are frequently asked to communicate better with members of Congress and with the public in general. The six letters that follow attempt to do just this. I am writing about human diseases we are trying to prevent and cure, and I wish to make three points which I shall illustrate with examples.

The first point is not fully appreciated by non-scientists. Despite remarkable advances made in the last three decades, our fundamental knowledge in many fields of health sciences is still rudimentary. Without this fundamental knowledge we are blindfolded and confused. I shall show how basic science can help us advance on several medical fronts and how a discovery in one area may bring progress and success in another, unrelated, field.

Second, I offer proposals on how basic science should be supported. Scientists have been asked to give such advice, and members of Congress need this advice to make the practical decisions the public has entrusted into their hands.

Third, we have to weigh the benefits derived from science against the "evils." I shall demonstrate how some of the "evils" can be avoided by better science and by more cautious use of technological advances.

In the *New York Times* item, reprinted above, Senator Kennedy characterized the research priorities of the scientists at the National Institutes of Health as "waiting for lightning to strike." I rather prefer another analogy, drawn by Dr. A. M. Weinberg at a conference sponsored by the Subcommittee on Government Research. He quotes his high school basketball coach: "In setting up a good shot at the basket, by all means keep the ball moving. It doesn't matter so much where the ball moves as long as it does not remain in one place; only in this way are openings created." Weinberg continues: "This approach to basketball is certainly inefficient. . . . yet by following this prescription our team won most of its games. In the same sense science is inefficient; by maintaining scientific activity in areas that are broadly of interest, one creates opportunities that can be exploited practically."

The scientists at the National Institutes of Health are not waiting; they keep the ball moving at a fast rate. They have created the most impressive center of medical research in the world and have netted many balls.

During the hearings referred to in the *New York Times*, Senator Kennedy angrily protested that the mission of the National Institutes of Health is not the understanding but the "prevention, diagnosis and treatment" of diseases. Is it possible he believes these aims can be achieved without understanding of the disease?

I hope to convince members of Congress that, in the case of most diseases, basic research is the only logical way to attack the problem and that by choosing this route scientists are not evading their responsibilities to the taxpayers who support their research.

Above all, I hope to make a case that stable support of basic research, perhaps less generously than some scientists would hope but according to a firmer and less fluctuating formula than we now have, will be in the best interests of the people of the United States and elsewhere.

Efraim Racker
Cornell University
March 1979

Acknowledgments

I wish to thank many friends who made valuable suggestions for revision. Among them are Bernie Davis, Dale Corson, Philip Handler, Hope H. Punnett, Arthur Kornberg, Robert Morison, Dorothy Nelkin, Gottfried Schatz, and Lewis Thomas. When more than one of them recommended a change, I always responded; when only one of them raised an objection, I often responded; when two gave opposing advice (which happened only rarely), I left it the way it was. Thanks are due to Edward Tenner and Christine K. Ivusic of Princeton University Press who helped with encouragement and English grammar. Judy Caveney, my secretary, deserves a medal for retyping the letters over and over again.

Prologue

There are two approaches to the cure of diseases: basic research and applied research. These are poor labels because they mean different things to different people.

There is basic research unrelated to the solution of a clinical problem, basic research related to the solution of a clinical problem, clinical research unrelated to the cure of a disease, and clinical research related to the cure of a disease.

We need to support them all and not worry about classification, and we have to face the problem of allocation of funds. I have made a number of suggestions to meet this serious problem in Letter 5, but it essentially comes down to a simple formula. Let us make sure that a bright, young scientist who is excited about a problem in neurobiology and behavior, for example, can go ahead with a project and not have to switch to cancer research because that is where the money is. Let us make sure that a bright, young scientist can proceed with a brilliant project on the behavioral responses of bacteria to attractants and repellents without being ridiculed. Let us make sure that a young Benjamin Franklin of today can study a problem of electricity before knowing that one day the study will be useful for the development of diagnostic electrocardiography.

In a penetrating study, "Scientific Basis for the Support of Biomedical Science," published in *Science* (192, 1976), J. H. Comroe and R. D. Dripps analyzed, with the help of 140 consultants, the chronological events in the development of electrocardiography and other important clinical procedures, such as open heart surgery, treatment of hypertension, and cardiac pacemakers. How could anesthesia have been discovered without the chemists who synthesized ether and chloroform? How could open heart surgery be performed without anesthesia? How could cardiac defects be diagnosed with a realistic prognosis of repair by surgery without selective angiocardiography, a method which rests on the earlier discovery of cardiac catheterization, which depends on X-rays? Did William Roentgen work on heart diseases when he discovered X-rays? No, he was curious to find out how electricity behaved in a vacuum.

Most members of Congress are in favor of basic research. They have said so repeatedly. But they show signs of impatience. Why can we not speed up the cure of cancer, diabetes, or multiple sclerosis when we can land a man on the moon? Why don't we know more about simple things like food—what's good for us, what's bad for us? We need to remind them that food is life—life of a cow, life of a plant—and life is complex, much more complex than a trip to the moon, much more mysterious, and much more challenging. We don't know much about cancer because we don't know much about how cells grow, divide, and differentiate.

We need to know more about normal cells and we need to understand the causes that lead to diseases. We have many clues and many gifted young scientists are excited about analyzing these problems. Let us make sure that they will be able to do so. This is the basic message of this book.

Science and
the Cure of Diseases

*Letters to Members
of Congress*

Science and Mental Diseases

My first letter starts with mental diseases because this is where my research as a medical student began. My primary interest then was the cure of mental diseases. How did it come about that now, some forty-five years later, I am doing basic research on membranes?

Wagner-Jauregg, a famous Viennese psychiatrist who received the Nobel Prize in 1927 for his treatment of patients suffering from syphilitic general paresis *dementia paralytica*, said in his first lecture to medical students in Vienna: There are two kinds of psychiatrists. First there are those who came to the profession with conviction and devotion, who knew from the first semester of medical school that psychiatry was their field of interest. They are fascinated by mental diseases and not afraid to confront the violence and confusion of the mind. They want to cure psychotics and neurotics, they are psychiatrists at heart. Then there are the other types of psychiatrists who got into the field by accident, because they really didn't know what else to do. They didn't want to get up in the middle of the night to deliver babies or they happened to be offered a well-paying job in a state mental hospital. "I want you to know," said Wagner-Jauregg, "that I belong to the second category." He was very suspicious of the first category and coined the phrase that the major difference between the psychiatrists and the patients in a mental hospital is that the psychiatrists had a key.

I must confess that I belonged to the first category. I

wanted to work with patients with mental disorders and I wanted to have a key—a key to the building but also a key to the understanding of the deranged mind. I was impressed with the relevance of the problem, the enormity of the economic burden to society which has to take financial responsibility for millions of mentally sick people. But most of all, I wanted to help these patients out of their nightmares.

Having been raised in Vienna to the music of psychoanalysis and the lullaby of the Oedipus complex, I turned first to the psychoanalytical approach, but I was soon plagued by doubts whether psychoses could be approached by psychoanalysis. In fact, Freud believed in the organic basis of the genesis of psychosis. He said in his book *Beyond the Pleasure Principle* (New York, 1924): "The shortcomings of our description would probably disappear if for the psychological terms we could substitute physiological or chemical ones." He continued: "Biology is truly a realm of limitless possibilities; we have the most surprising revelations to expect from it, and cannot conjecture what answers it will offer in some decades to the questions we have put to it. Perhaps they may be such as to overthrow the whole artificial structure of hypotheses."

In 1938, an invasion of a mass psychosis took place from Germany to Austria, and I hurriedly left for England. There I came across an article by J. H. Quastel, a brilliant British biochemist, on "Biochemistry and Mental Disorder" (in the book *Perspectives in Biochemistry*, Cambridge, England, 1936). I went to work with Dr. Quastel in a state mental hospital in Cardiff, Wales, on the effect of drugs on brain metabolism.

Quastel's hypothesis of mental diseases was this: The proteins of our food are digested in the body to amino acids, which are the building blocks of proteins. By splitting off one molecule of carbon dioxide, amino acids are then metabolized to toxic substances called amines. If there is something wrong with either the rate of produc-

4

tion or the removal of these amines, brain metabolism is affected and mental disease may arise. There were two experimental facts on which this hypothesis was based. One, certain amines when given in large doses cause psychosis-like conditions. I mention three such drugs which are widely known: Mescaline, LSD, and amphetamine. LSD can precipitate a psychosis, and the chronic use of amphetamine may cause a clinical picture remarkably similar to schizophrenia. Two, Quastel found that various amines inhibit the metabolism of rat brain slices in the test tube.

Quastel's idea was that since sugar is the major food of the brain, we should look for changes in sugar metabolism induced by toxic amines. One of the experiments I did in Cardiff, with all the optimism and enthusiasm of youth, was to expose rat brains to the blood of schizophrenic patients. I reasoned that if toxic amines were present in their blood, they should inhibit the sugar metabolism of the rat brain. These experiments were doomed to failure because I knew too little about sugar metabolism and about the complexity of brain function and structure. But above all I knew too little about biogenic amines, so-called, because they are normal and biologically important ingredients of our body. Among them are potent hormones such as dopamine, adrenalin, and serotonin. As we shall see, these biogenic amines regulate our emotions and alertness. Many psychiatrists now believe that a disturbance in the function of biogenic amines in the brain is a key feature in psychosis. Amphetamine and LSD are toxic amines believed to induce psychosis by interfering with the proper function of the natural biogenic amines.

When I immigrated to the United States in 1941, I encountered no interest in either toxic or biogenic amines, but there were funds available through the March of Dimes for research in poliomyelitis. I was happy to get a job with a salary of 12,000 dimes per year to explore the effect of polio viruses on brain metabolism. I soon discovered a defect in the utilization of sugar in the brains of mice infected with polio virus. But once again I could not proceed logically with the problem because we were lacking fundamental information on sugar metabolism.

This was a turning point in my career, for I realized that without fundamental knowledge of the biochemical processes we cannot understand diseases of either the body or the mind; we cannot design a logical approach to either their treatment or their prevention. I therefore turned to the study of basic processes of carbohydrate and energy metabolism.

Diagnosis and Classification

During the first decades of this century, psychiatrists developed the diagnosis of mental diseases—an important contribution. To make a diagnosis we need a classification, and the most useful classifications are dependent on basic studies, either biological or psychological. If we have a patient with an insulin deficiency, we treat the patient with insulin. However, if we were to call every patient with sugar in the urine a diabetic, we would find many patients who would not be helped by insulin. The trouble with most mental diseases is that we do not know their basic causes and we diagnose by symptoms and signs only. Yet the symptoms are even more vague than sugar in the urine. It is therefore remarkable that the classification of psychoses into two major groups—schizophrenic and manic-depressive diseases—turned out to be valuable and consistent with later genetic studies. However, in 1911, Bleuler in Switzerland recognized the complexity of schizophrenia and referred to "groups of schizophrenia." In 1966, under the sponsorship of the World Health Organization, an international classification of mental diseases was adopted. It resembles other international disagreements. In the preface of the official booklet of the American Psychiatric Association on the classification of mental diseases, the following comment was made regarding schizophrenia and the deliberations of the international committee. "Even if it had tried, the committee could not establish agreement about what this disorder is; it could only agree on what to call it." In the international classification, schizophrenia has the number 295: 295.0 is the simple type; 295.2 is the catatonic type; 295.23 is schizophrenia, catatonic type, excited; 295.24 is schizophrenia, catatonic type, withdrawn; and 295.90 is schizophrenia, chronic undifferentiated type. As can be seen, the classification is based purely on signs and symptoms. What is badly needed in psychiatry is a more biologically oriented classification. If we could find out why some schizophrenics are catatonic and why some are not, why some patients are always depressed while others go through cycles of depression and excitement, we might be able to refine our classifications. Responsiveness to drugs may also help us in improving the classification we now have.

Genetic Factors in Mental Diseases

The important role of genetic factors in mental diseases is now widely accepted. Well-controlled studies with children of psychotic patients adopted shortly after birth established not only a genetic role in psychoses, but supported the psychiatric classification. Schizophrenics are more likely born to schizophrenics, and manic-depressives to manic-depressives. But it is not a simple problem, and the most favored view is that there are multiple genetic and environmental factors that influence susceptibility to these diseases. These various factors need to be elucidated, and we should avoid a fatalism often associated in the public mind with the verdict of a genetic disease. There are no diseases, including infections, that are not influenced by genetic factors. As I shall elaborate, even diseases caused directly by an altered gene are amenable to a therapeutic approach.

In this connection, the mental disease phenylpyruvic oligophrenia, or phenylketonuria (PKU), is particularly important because it has influenced psychiatric thinking and has removed some of the stigma of a "genetic" disease. Here is a biochemical disease with a known lesion. An enzyme which metabolizes phenylalanine, one of the natural amino acids in our body, is lacking, and PKU patients have high blood levels of phenylalanine. Since phenylalanine is metabolized to dopamine, the relation to the biogenic amines becomes obvious. What is the cause of the mental retardation of PKU patients? It appears from animal studies and recently also from clinical patients that high blood levels of phenylalanine are very toxic in early life but much less so in adulthood. Therefore, mental retardation can be prevented, or at least alleviated, in many cases by giving PKU babies a diet low in phenylalanine.* This must be done early, and fortunately in many states testing of newborn babies for PKU is compulsory. There are now many cases of PKU adults who had been successfully treated with the diet shortly after birth.

PKU is a mental deficiency disease and not a psychosis. Another, much rarer, genetic disease that gives rise to psychotic symptoms is Hartnup disease, which is associated with the faulty metabolism of another natural amino acid called tryptophan. The biogenic amine serotonin is manufactured in the body from tryptophan. Thus we see another possible link between psychoses and biogenic amines. But tryptophan is also required for the production of the vitamin nicotinamide, a representative of the vitamin B group. A lack of nicotinamide gives rise to a disease called pellagra, which has, among its symptoms,

---

* A low phenylalanine diet (e.g. Lofenalac®) consists of about 50 percent corn syrup solids, 18 percent degraded casein (with low phenylalanine content), 18 percent starch, and salts, amino acids, and vitamins. It is important to check phenylalanine blood levels of the patient to make sure they are not too low for growth and not too high to damage the brain.

8

lesions of the skin as well as psychotic manifestations. The faulty metabolism of tryptophan in Hartnup disease may lead to deficiency of a vitamin as well as of a biogenic amine. We are only beginning to learn about the complexity of interactions between such multiple factors. Moreover we now know that a deficiency in one biogenic amine or in one vitamin influences the functions of other amines and vitamins. We need to know more about these fine-tuning mechanisms in general and more about the specific changes in biogenic amines and vitamins in Hartnup disease before we can design a rational treatment.

An intriguing genetic disease was described by M. Lesch and W. L. Nyhan in the *Journal of the American Medical Association* (36, 1964). Children with this disease have symptoms of mental retardation, spasticity, strong tendencies of aggression, and compulsive self-mutilation. Unless strapped, they bite their hands and arms, severely wounding themselves. The genetics of this disease is well established. There is a defect in one gene which results in the lack of a single enzyme. How this single metabolic defect gives rise to a tendency of self-mutilation and aggression is completely unknown.

Biochemical Factors and Drug Therapy

What about the more frequent manic-depressive diseases and schizophrenia? Are they also biochemical diseases or are they environmental? Even though environmental factors play a role, I am convinced they are basically biochemical diseases because we can detect biochemical changes, because psychoses can be induced by chemicals, because of genetic studies, and because very specific drugs affect the disease. Although this kind of evidence is very circumstantial, I subscribe to this opinion because the biochemical hypothesis is the one which gives us hope for a cure.

Drugs such as chlorpromazine have radically changed

both the clinical and sociological picture of schizophrenia. A few patients respond so well that they can leave the hospital without further treatment. Some have to be kept on the drug and some do not respond at all. In economic terms, the most important fact is that many patients become well enough to leave the hospital while remaining members of a therapeutic community.

The history of chlorpromazine is a fascinating story that should be known to every administrator responsible for the allocation of research funds, for it shows how completely unrelated research projects may profoundly influence another field. During the middle of the nineteenth century a chemist, while searching for a cure for malaria, accidentally prepared the first synthetic dye. This started the synthetic dye industry which is largely responsible for the colorful display of today's textiles, plastics, and paints. The dye industry more than once repaid their debt to medicine. Methylene blue was among the thousands of dyes that were synthesized. A curious cycle of discoveries was closed when, in 1891, Paul Ehrlich found that methylene blue is an effective drug in the treatment of malaria. Research in France during World War II led to the discovery that some compounds related to methylene blue were not only effective against malaria but counteracted certain allergic reactions. Again, by accident, it was discovered that some of these phenothiazine drugs, including chlorpromazine, made anesthetics more effective. It became an aid in the treatment of surgical shock and was found to have a calming effect on the behavior of patients. Chlorpromazine was finally tried on a mentally disturbed patient, and medical history was made. What a tortured road from malaria, via synthetic dyes, back to malaria, then to allergy, surgery, and schizophrenia!

Is there an administrator or scientist who could have stood up before a congressional committee and defended the study of methylene blue, anaphylaxis, the mode of action of histamine, or surgical shock as relevant for psy-

10

chiatry? Yet the Evaluation Policy Committee of the National Institute of Mental Health has selected work on phenothiazine drugs as the most outstanding contribution in the past two decades in the treatment of schizophrenia.

How much do we know about the mode of action of chlorpromazine? Or shouldn't we care? Should we wait another hundred years for a scientist to synthesize a chemical compound that perhaps will be tried out as an anticancer drug and found to have behavioral action? Or perhaps someone will try this new chemical as a potential antipsychotic drug in a cancer patient and cure the cancer. Let us hope that such accidental discoveries will continue to bless us, but should we wait for them? Is it not more logical to find out how chlorpromazine works by supporting basic research on its mode of action? Unfortunately, the problem is not simple and may sound very confusing to a non-scientist.

Chlorpromazine has many metabolic effects; for example, it inhibits an enzyme which adds a chemical group (a methyl group) to biogenic amines. This lends support to investigators who believe in the altered metabolism of biogenic amines in psychoses. Chlorpromazine also inhibits absorption of drugs from the gut. This fits with an old idea that intestinal intoxication is a factor in psychosis. Chlorpromazine inhibits the carbohydrate metabolism of brain slices; this fits well with the hypothesis of Quastel. Chlorpromazine has a toxic effect on membranes in general, and that pleases those who believe psychoses are membrane diseases. But the most striking effect of chlorpromazine is on the "receptor site" for dopamine in membranes where it interacts at remarkably low concentrations that are consistent with therapeutic levels of the drug used in the treatment of schizophrenia. Moreover, there is a good correlation between the therapeutic effectiveness of various phenothiazine drugs and their potency in the interaction with the dopamine receptors. This

11

opens up an exciting approach to the understanding of the disease. More than that, if this relationship is correct, it gives us a quick and inexpensive test-tube test for the evaluation of thousands of drugs that would take decades in clinical trial.

Another important contribution to psychiatry is the successful treatment of manic-depressive diseases with lithium. Lithium is a simple atom like sodium or potassium. How does lithium work? Again, there is no simple answer because lithium influences several metabolic systems. At present the most likely clue is that lithium affects the function of some biogenic amines.

A third type of therapeutic compounds effectively used in psychiatry, particularly against depression, are the so-called tricyclic antidepressants. They do not interfere with the level of biogenic amines, but they alter their distribution in the brain by interfering with their movement from one place to another.

Another drug that has helped to establish the role of biogenic amines is reserpine. It depletes the brain of biogenic amines. If administered to mice it produces sedation. In about 5 percent of patients with hypertension who are treated with the antihypertension drug reserpine, a state of depression resembling a psychosis is produced, and treatment with this drug must be abandoned. It may be very interesting to explore why this group of patients is so hypersensitive to reserpine.

Several metabolic derivatives of biogenic amines have been reported to be present in abnormal levels in patients with depression, suggesting a direct correlation between biogenic amines and this disease.

More about Biogenic Amines

It appears that three major drugs—chlorpromazine, lithium, and the tricyclic antidepressants—used in the treatment of psychoses affect biogenic amines. How much do we know about these remarkable compounds? How are they synthesized and broken down in our body? How are their levels in the brain controlled if they have such pronounced effects on our emotions, alertness, and motor activities? The biological catalyst MAO oxidizes amines, including biogenic amines, and thereby exerts a control on their presence in the brain. Inhibitors of this enzyme have been widely used for the treatment of mentally depressed patients. The story of the MAO (monoamine oxidase) inhibitors is rather sad. In fact, it is a good illustration of how a "practical" approach without sufficient basic background can delay useful progress. Some of the best MAO inhibitors were developed in the United States, but today drug companies are not interested in further developments in this area. The reason is that several deaths were reported among patients treated with MAO inhibitors. MAO not only regulates the level of biogenic amines in the brain but is present also in the liver, where it detoxifies some of the highly toxic amines that we ingest with our food or that are produced by bacteria in our intestine. For example, cheeses and wines are rich in tyramine, a compound that raises the blood pressure. The MAO inhibitors suppress this detoxification function of the liver, and patients who could not detoxify high doses of ingested tyramine died of cerebral accidents caused by high blood pressure. Now few physicians in the United States dare to prescribe these drugs, which are still widely used in Europe.

One day it occurred to me that there may be some fundamental differences between the amine oxidase enzymes in the liver and in the brain. Perhaps we could take advantages of such differences and develop drugs that would inhibit the brain enzyme without seriously interfering with the detoxification processes in the liver. When I failed to convince research directors in certain drug companies to undertake such research, I started a project of my own. Our research soon revealed that the brain con-

13

tains two types of amine oxidase enzymes. One is like the liver enzyme, but we found a second one which is particularly concerned with the metabolism of biogenic amines. We also discovered that one drug called harmaline inhibited over 90 percent of the brain enzyme activity without any effect on the liver enzyme. Harmaline is a bad drug—it causes hallucinations. But we established with our experiments that the brain enzyme is sufficiently different from the liver enzyme to justify a search for a suitable drug that would inhibit the brain type enzyme without interfering with the detoxification process in the liver.

My story illustrates how rushing into practical application before sufficient basic knowledge is available may actually delay the development of a useful drug. The drug companies are still too scared by the bad publicity for MAO inhibitors to pursue the problem further.

New brain hormones are presently being discovered. Among them are some interesting compounds called enkephalins that act in some ways like opium. It has been reported that these agents specifically affect the level of dopamine in the brain. Speculations on a possible role of enkephalins in schizophrenia have been published and publicized.

Other queer compounds are synthesized by the brain such as dimethyltryptamine, a hallucinogen known among the younger generation under the psychedelic name of "ultimate spinach." Nothing is as yet known about its normal physiological function.

Scientists are wary, and should be wary, about making promises and holding out hopes for cures of this or that disease. There is an arduous and long journey before a promising compound in the laboratory is finally put to clinical use. Many compounds do not make it. But we should realize that the first and most important steps in the treatment of mental diseases have been taken. Tricyclic antidepressants, chlorpromazine, and lithium are

14

important drugs not only because they help but because they have lifted the defeatist attitude that has characterized psychiatry of the pre-chlorpromazine era. On the other hand, we must realize that chlorpromazine and lithium are far from ideal drugs. Some people have severe side reactions; some patients do not respond. Drugs with higher potencies, less side reaction, and a wider spectrum of effectiveness are needed. We can expect that a deeper understanding of how they work will allow us to design better ones.

Where Do We Go from Here?

There is no other field in which the need for basic research is more persuasive. Because of the overwhelming complexity of the brain, relevant research may be on computer simulation, on membranes, on the mechanism of enzyme action, or on microchemistry suitable for the analysis of single brain cells. As in the case of chlorpromazine, progress in the treatment of mental diseases may come from an unexpected direction.

If there is no limit to the number of relevant problems, how shall we allocate funds? I shall talk about this problem in greater detail later, but it seems to me that the basic solution is a simple one. Let us make certain that no bright, young investigators who want to devote their research life to the study of neurobiology shall become frustrated because they cannot obtain support for their work. Let us suppose he or she has the idea that the best approach to the study of the sensory system is to analyze the curious response of bacteria to chemicals that either attract or repel them. Is this any less related to mental diseases than the studies of Benjamin Franklin on lightning were related to the diseases of the heart and relevant to the development of the electrocardiogram?

Many young biologists are attracted to neurobiology because of the challenge of the unknown and the relevance of the field to human behavior. Some of these young people have come to me for advice. They have asked me if it would not be wiser to take up a project on cancer research because it is more likely that it would be supported, while a grant application on the "swimming properties of bacteria" may be turned down. Or if funded, it may be challenged by a senator who has ridiculed studies on the "social behavior of the Alaskan brown bear" or on "environmental and physiological causes of aggression" supported by the National Science Foundation. How would the senator have reacted to the application of Gregory Mendel, the father of genetics, entitled "How to segregate round and wrinkled peas," or of T. H. Morgan, another pioneer in genetics, "On the sex life of the white-eyed fruit fly"? Do these titles sound any more promising than the "Study of the use of Qat [a drug similar to benzedrine] by the people of North Yemen"? I do not know whether any of the studies ridiculed by the senator should or should not have been supported by the National Science Foundation. I do not know whether the plans for these studies were scientifically sound or whether the experiments have been well executed. This is for the experts to decide. Is it fair to sway public opinion because the titles sound irrelevant? I believe that one of the challenged studies by a brilliant ecologist on the love life of some African birds, which reveals a fascinating pattern of social behavior, may be more relevant to the study of mental depression that a crash program on melancholic patients with a thousand drugs.

Our young scientists are excited about the fields of neurobiology and behavior. Let us preserve this precious excitement and allow them to use their imaginations and study the problem of their choice.

Science and Cancer

Cancer is a complex disease. Although there is one common denominator, namely *uncontrolled cell growth*, there are many forms of cancer and many contributing factors. We know that viruses as well as chemicals (called carcinogens) can cause cancer in man and in animals.

Viruses as Cancer-Causing Agents

It is fashionable now to talk about the virus etiology of cancer, but I remember that only thirty years ago a mere handful of scientists believed that viruses may play a role in human cancer. Among them was Colin MacLeod, the much loved and admired head of the microbiology department at the New York University School of Medicine. He induced me to teach a laboratory course on viruses with particular emphasis on their role in cancer diseases of animals. I believe it was the first course of its kind, and I taught it with enthusiasm for three years until I left for Yale University in 1952. I mention this course because I have been convinced for a long time that viruses play an important role in the etiology of cancer. But I am equally convinced that the current fashion has gone too far, and large financial support has been given to attempts to isolate "the human cancer virus." These projects have resulted in premature claims, disappointments, and discords. They have contributed more to daily newspapers than to the progress of cancer research. I do not want to discourage the search for human cancer viruses, but we should look at this approach with the proper perspective without raising hopes that such a discovery will bring us closer to a cure of cancer. Above all, we should only fund those efforts which are carried out by highly competent scientists or we will do more damage than good. It is now more than half a century since convincing evidence emerged that viruses can cause cancer in animals. In spite of this knowledge we cannot cure these afflicted animals and we still have no answer to the fundamental question of why and how a virus causes cancer.

Why can't we prevent cancer if it is caused by a virus just as we did with polio? There are at least two reasons. We know of dozens of viruses (and there may be hundreds) that cause cancers in animals. To prepare an effective vaccine that will induce production of specific antibodies in our blood, we *must know and must use* the specific virus or a closely related one that causes the disease. This is a serious drawback of vaccination procedures in general and is under lively discussion in connection with the current problem of influenza vaccines. We are lucky that there are only three polio viruses which can attack us.

Another and equally serious problem is a curious phenomenon of cancer biology, which was discovered by Richard Shope at the Rockefeller Institute over thirty years ago. He showed that we can indeed prevent the growth of a tumor in an animal if we supply antibodies *before* we infect it with the virus. The antibodies prevent the entry of the virus into the cell. But once the cell is infected, it becomes a tumor cell, grows, and divides, and the antibodies sit helplessly outside the cells they cannot enter. Thus the virus eventually kills its "host," the name used in the gentle language of the virologists for the animal that has become infected.

18

Chemicals as Cancer-Causing Agents

There are now several thousand chemicals known to give rise to cancer in animals. The first potent carcinogen was isolated from coal tar, which was known to be a cancer-causing agent. The isolated compound is called benzo(α)pyrene. It is a simple chemical containing twenty carbon atoms arranged in five benzene rings:

STRUCTURE OF BENZO [α] PYRENE

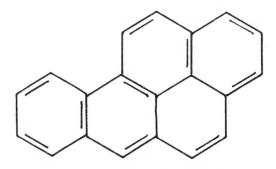

Benzo(α)pyrene is the main carcinogen in cigarette smoke, but it is present in many other products—e.g., oils and fats that were exposed to high temperatures. That includes a wide range of sources from exhaust fumes of internal combustion engines to charcoal broiled steak and french fries. Pollution of our waters by the waste products from industrial plants include benzo(α)pyrene and many other related carcinogens that may eventually contaminate our drinking water.

When should we start to worry about a carcinogen? When does a carcinogen become dangerous? Let me describe an experiment before I deal with this important and complex question. If we expose the skin of a mouse for about two months to half a milligram of benzo(α)pyrene per week, a tumor will appear. If we then stop the application, the tumor will regress and disappear in most mice. But if we continue to apply benzo(α)pyrene for another month or two the tumor will become malignant and kill most of the mice. Can we take this as a model for human cancer? I think we should in view of our present knowledge about cigarette smoking and lung cancer. Four months in the life span of a mouse is more than a decade in the life span of man; a total of four milligrams of benzo(α)pyrene is a lot of carcinogen for a little mouse. If we give less, it takes longer to develop the tumor. These experiments explain a great deal about cancer in men. With the dramatic increase in life expectancy during the past hundred years, there is also an increased probability of getting cancer because we are continuously exposed to carcinogens. These carcinogens are not only manmade chemicals put into our environment and food. Often enough they are ingredients of so-called natural food. Some plants manufacture carcinogens, others may be infected with molds that are potent producers of carcinogens. One of them, aflatoxin, is one of the most potent carcinogens known and is found in peanuts and other plants infected with molds. Intestinal bacteria can generate nitrosamines that are potent chemicals that can cause cancer in almost every organ of our body. Ultraviolet light and X-rays can also contribute to the development of cancer.

The question is not whether we are exposed to carcinogens but how often and how much is required for the appearance of cancer cells. We do not know the answer to this question. We do not know enough about the relative susceptibility of, e.g., mice and men to the same carcinogen. We do not know enough about simultaneous exposure to multiple carcinogens. We do not know how many carcinogens help each other in causing tumors or how many promoters there are or how they work. Promoters are chemicals that do not cause cancer by themselves but

20

enhance the effectiveness of carcinogens. We need to know more.

Meanwhile, how should we respond to the fears stirred up by reports that stringbeans, or saccharine, or charcoal broiled steak may be either carcinogenic or may be promoters. The best answer I know, particularly for the population of the United States, is to eat less and more varied food, thereby reducing the probability of excessive exposure to carcinogens in the food. By doing so, we shall improve our life expectancy, which is lower for fat people than for thin people. Obviously, we must also try to eliminate carcinogens from our environment.

Relevance and Fashion in Cancer Research

Can we find a cure for cancer quickly if we put a lot of money into cancer research? Former President Nixon, inspired by the success of the "Man on the Moon Project," called for an "attack on cancer," asked Congress for funds for expanded research, and signed into law the National Cancer Act of 1971. During the winter of 1971-1972, two hundred and fifty "prominent laboratory and clinical scientists" met at the Airlie House Conference Center in Warrenton, Virginia, to develop the foundation for a National Cancer Program. I was among those invited. In spite of serious reservations, which I shall explain, I agreed to participate in the sessions. I came home from this conference with a severe headache. Since I get headaches very rarely, it is not too surprising that I remember coming back from another conference about ten years earlier with a similar headache. This was also a conference on cancer—a review session for the American Cancer Society.

The earlier incident is pertinent to my evaluation of the Airlie Conference and the National Cancer Program. Many years ago, the American Cancer Society had installed an effective peer review system to deal with the applications for funds in support of cancer research. Originally, the members of these peer review panels had been instructed to evaluate the scientific merit of the proposals in the tradition of the committees organized by the National Institutes of Health. But at that meeting a new director of the American Cancer Society informed us that he had decided that the review system needed to be changed. He requested that the scientists evaluate, with a priority rating of one to five, two aspects of the proposal—the scientific merit and the relevance to cancer research. He was prepared with printed forms featuring the two evaluation columns. I argued vehemently against this procedure and refused to go along with it. My point was simple enough. I said that I did not know what kind of biological research is *not* relevant to cancer. Does anyone? I remember distinctly the grant application which we discussed when my headache started. It was a project on the isolation of various plant products and their effect on photosynthesis. Clearly, the director said, this is not relevant to cancer and we should give it a priority of five for relevance. If the science is excellent we should give it a priority of one for scientific merit. He and his staff will have to decide which grant should be paid and which should be refused. It was not hard to predict the fate of this application.

Can a research project on photosynthesis be relevant for cancer? I believe it could be, and I shall relate later how our own work on cancer has been helped by studies with plant products. Perhaps more persuasive is the fact that the National Cancer Institute now has an active program on the isolation and characterization of plant products that have promising antitumor activity. What I want to emphasize here is that the director of the American Cancer Society was not justified in his statement that a study on plant products and photosynthesis is clearly irrelevant to the cancer problem. I hasten to add that this director is no longer with the American Cancer Society,

22

which currently supports many excellent basic as well as applied research projects.

I hope that my report of this earlier experience explains my serious reservation about my participation in the Airlie House Conference. There was another equally disturbing reason. I was concerned about assigning a great deal of money to a crash program, supporting work that is fashionable rather than first rate, raising hopes that we cannot sustain. Since the resources of funds for research are limited, the funding of "relevant" or fashionable programs that are not necessarily of high caliber may divert money from research projects of superior scientific merit.

My fears about the Airlie House Conference were justified. Many scientists felt the pressures of relevance and fashions and some submitted to them. In my opinion, the National Research Program Strategy Hierarchy that has emerged from these meetings is a ludicrous document that will have no impact whatever on the progress of cancer research. I do not know how many millions of dollars were spent to develop this program, but I believe it was virtually all wasted. Fortunately, the extra money provided by Congress for cancer research was not wasted. Why? Because of the wisdom of individuals on the President's Cancer Panel, the National Cancer Advisory Boards, and the National Cancer Institute, who have a deep understanding of the significance of basic research and have used good judgment and given good advice in the distribution of funds. But despite their goodwill and understanding, some of the administrators at the National Institutes of Health had to yield to pressures from Congress, particularly in the case of the so-called contracts. I have served for several years as a member and chairman of one of the advisory boards to the National Cancer Institute. My colleagues and I have watched with dismay the disproportionate assignment of money to these contracts, which often include projects that do not meet the high scientific standards applied in the review of project type research grant applications. I realize fully the need for some contracts aimed at practical problems. I realize also the difficulties that members of Congress face in making decisions of this type—how much money for project grants, how much for contracts? They need better guidelines, and I shall return to this problem in the fifth letter of this series.

What Research Is Needed in the Cancer Field?

Prevention of cancer. I outlined earlier the difficulties we face with respect to a possible vaccine against cancer-causing viruses. I shall therefore concentrate on cancers caused by chemicals.

We need to know more about how chemicals cause cancer. We need to know much more about additive effects on exposure to multiple carcinogens. We need to know more about what happens to carcinogens in our body and how some carcinogens are detoxified and other innocuous chemicals are converted in our liver or intestine into potent carcinogens. We need more information on genetic predisposition to cancer and on the effect of hormones on susceptibility. These are basic problems that require much more fundamental knowledge of normal growth processes than we presently have.

We need more and better epidemiological studies. We have many clues that need to be pursued. I do not have to repeat the well-known story of cigarette smoking and lung cancer. I would like to record here, however, a viewpoint which I do not share. In an otherwise very valuable article published in November 1975 in *Scientific American*, the British biologist Dr. John Cairns wrote: "It could even be argued that few Western societies could afford to abolish a habit that creates a large secondary industry, generates considerable revenues and kills mostly the older members of the population who otherwise would draw on government welfare and social security benefits."

24

I assume Dr. Cairns has written this with tongue in cheek, but the view is cynical and I believe incorrect. We *can* afford it, and I challenge even the soundness of the budgetary argument that does not consider the cost of prolonged hospitalization and treatment which many governments are now providing for cancer patients, not to speak of the anguish and pain of the chronically ill.

Lung, intestine, stomach, and breast are the most commonly affected organs, accounting for about half of all cancer cases. There is evidence that in all four categories environmental factors play a role. Death from stomach cancer in the United States has dropped precipitously in the past fifty years. We do not know why. In Japan the incidence is much higher, but offspring of Japanese who have immigrated to the United States lose the predilection for stomach cancer. Intestinal cancer is more frequent in rich countries, and there is considerable evidence on the stimulating effect of a high caloric diet on cancer in experimental animals. There is some suggestive evidence collected by Dr. Burkitt at Johns Hopkins University and others that a high residue diet (e.g., high in cereals and low in beef) accelerates the movement of the bowels, decreasing the incidence of intestinal cancer. Dr. Cairns has documented in his *Scientific American* article an impressive correlation from epidemiological studies in over twenty different countries between the incidence of intestinal cancer and a high meat (low cereal) diet.

We need quicker and cheaper assays for the evaluation of carcinogens. Over 100,000 new organic compounds are being synthesized each year. Before any of them can be put to use that involves exposure of human beings, they need to be tested for toxicity and carcinogenicity. Unfortunately, assays for carcinogens in mammals, the best tests we have, are very expensive and time-consuming, even when mice or rats are being used. Moreover, the limitations of such assays are quite obvious. Only a positive test has much meaning because a compound may be carcinogenic in men but not in mice or rats. But the argument can and has been turned around; a carcinogen in mice need not be a carcinogen in men. To be more meaningful, the test must be performed in several animal species and extended to about half of a life span of the animal (1 to 1.5 years in rodents).

Several assays that are cheaper and more rapid have been developed. Most extensively used is the Ames test. Dr. Bruce Ames, of the University of California, has developed a cheap and simple test to screen chemicals which several hundred industrial companies have now adopted. The test is controversial. Dr. Ames does not use mammalian cells, which get cancer; he uses bacteria which do not get cancer. He chose bacteria because we have the basic knowledge required for a meaningful test, knowledge that is much harder to get with mammalian cells. Dr. Ames does not measure tumor formation, but mutations—alterations in the genes. There is a great deal of evidence that mutagens (substances that increase mutations) are also carcinogens. The correlation between the two is not perfect, but very good, probably better than 90%. We might miss a few percent of the carcinogens and may include a few percent of compounds that are mutagens in bacteria but not carcinogens in mammals. Considering the difference between a bacterium and a mammalian cell, the correlation is truly remarkable and contains an important message which I shall discuss later. About fifteen years of hard basic research by Dr. Ames and many years spent by other investigators have made it possible to devise this simple test.

He uses a bacterium that has lost the ability to synthesize the amino acid histidine, which is needed for growth. In a medium lacking histidine this bacterial mutant does not grow; it is called a histidine-dependent mutant. If histidine-dependent bacteria are exposed to mutagens, the probability that a back-mutation occurs is greatly increased. Such a "revertant" has regained the ability to synthesize histidine and to grow in a medium which contains no histidine. The more effective a chemical is as a mutagen, the more back-mutations are observed.

In a laboratory course which Dr. Ames teaches, he once sent students to neighborhood pharmacies to buy cosmetics and various over-the-counter drugs for analysis of their mutagenic activities. One student brought back a hair dye which had extraordinarily high mutagenic activity. Ames then collected a large variety of hair dyes, and the tests revealed that almost all hair dyes sold for women were highly mutagenic. Hair dyes for men were of a different type and did not contain the mutagenic compounds used in dyes for women. The reason for this "discrimination" was revealed by one of the representatives of the hair dye industry. Hair dyes for women are more potent, but they darken the urine in individuals who use them. The observant man would quickly discover this, while dilution by the water in the bowl would obscure the color and deceive even an observant woman. The shocking aspect of this story is that these dyes must circulate in the bloodstream and may act as carcinogens before they are excreted into the urine.

We need to know more about genetics and mutations. Why is there such a close relationship between the mutagenic and the carcinogenic properties of drugs, a fact which has been known for many years? We know that the chance of getting cancer increases greatly with advancing age. With some types of cancer the annual death rate at the age of sixty is over ten times that at the age of forty. It has been suggested by several scientists that the striking relation to age and the mutagenic properties of carcinogens fit together if we assume that several genes have to be mutated before a patient has manifest cancer. We know from experimental research that multiple factors play a role even in a virus-induced tumor, such as the breast cancer of mice. Hormones, diet, as well as genetic factors, influence the course of this disease, which is transmitted by a virus in the mother's milk.

This idea that multiple mutations are required for the

manifestation of clinical cancer accounts not only for the pronounced increased probability with aging but it explains another very important phenomenon. Exposure of young people to carcinogens—e.g., by smoking or by X-ray treatment for acne—is much more dangerous than for old people because time is required for additional mutations to occur. It also tells us why stopping to smoke at an advanced age may not prevent the growth of a cancer in a lung that has already been primed for development.

Much more basic work is needed on the relationship between chemical structure and mutagenesis or carcinogenesis. It is important for people who get impatient with the slow progress of science to realize the complexity of biological problems that seem simple at first glance. For example, chemists who thought that they had some understanding of the chemistry of carcinogens were much surprised to learn of the carcinogenic potency of some compounds that did not fit the pattern. It turned out that these compounds themselves are indeed not carcinogens but are converted into carcinogens in the liver by a metabolic pathway which is normally used to degrade toxic compounds that get into our body. Dr. Ames was aware of this and included in his test an enzyme preparation from liver to make sure that he did not miss these convertible compounds. To complicate matters further, the liver enzymes which make carcinogens of some chemicals and detoxify many other compounds that may be carcinogenic are influenced by a variety of dietary and environmental factors (e.g., ingestion of barbiturates). These liver enzymes are now being studied in many outstanding laboratories all over the world and knowledge is slowly growing.

Treatment of Cancer

What can we do once we know an individual has cancer? What can we do to prevent the tumor from growing and spreading? What can we do to make it regress?

28

I shall not discuss current methods of treatment by surgery, radiation, and chemotherapy. But there is one phenomenon which is becoming increasingly apparent: Each cancer patient is different, and each tumor has an individual biology that we need to understand for proper treatment. There is a form of cancer of the testicle that is quite malignant. But if removed surgically and if the patient is treated for a month with X-rays, the probability of complete cure is better than 95 percent. Some tumors respond to radiation treatment, some to chemotherapy. Sometimes a single drug is effective, sometimes a combination of several drugs is needed.

I want to emphasize the difficulties we are facing in the evaluation of effective treatment. Some cancer patients, particularly those with certain leukemias, have remissions, sometimes months of apparent improvement, giving the illusion of cure. This is why claims are often made for the effectiveness of drugs or diets that are not scientifically proven. It requires a large number of cases and proper controls to establish the effectiveness of a drug. Individual testimonies, as impressive as they may sound, have really not much meaning.

I want to discuss now how future research can contribute to the treatment of cancer. First of all we need to know more about the biology of cancer. What are the reasons for the rapid growth and invasive properties of cancer cells? There is only one honest answer to this question. We do not know. But there are literally hundreds of clues that are being intensely explored by many outstanding investigators all over the world. Dr. Robert Holley received the 1968 Nobel Prize in medicine for his brilliant work at Cornell University on the chemical structure of "transfer ribonucleic acid," RNA, a complex compound which is involved in the synthesis of proteins. About twelve years ago, Dr. Holley changed fields. He wanted to work on a "relevant" problem and chose to study differences between normal cells and cancer cells grown in tissue cultures. It looked simple at first when he discovered a factor present in blood to which both cells responded differently—much less of the factor was needed for abundant growth of cancer cells. But over the years the picture has become increasingly complex. There is not one but several factors in the blood which affect the growth of cells, and it will take many years to identify and characterize them. It must be done, and progress is being made in numerous laboratories on the isolation and characterization of these various growth factors.

Meanwhile we can approach the same problem differently. Why do cancer cells and normal cells respond differently to growth factors? Are there some differences in the properties of their surface membrane? Indeed there are, and each difference we discover must be studied intensively and examined for its possible significance with respect to the phenomenon of malignancy. Immunologists and chemists are working intensively to isolate membrane components from normal cells and cancer cells. Among the rational approaches to the treatment of cancer, one is to take advantage of these differences in the cell membrane components either by designing specific chemical attacks or by taking advantage of defense mechanisms—e.g., those that involve white blood cells.

Rapidly growing cells such as cancer cells need to synthesize important cellular constituents called nucleic acids from specific building blocks in our food. On the basis of this knowledge, chemists have synthesized false building blocks that are mistaken by the body for the real ones and used for the synthesis of faulty nucleic acids that lead to the death of the cancer cells. These false building blocks (e.g., fluorouracil) are now widely used in the chemotherapy of cancer patients. Unfortunately, other cells that multiply rapidly, such as in the intestinal tract, are affected also and the drugs are often quite toxic. We have to learn more about the fundamental properties of these rapidly growing normal cells that are not cancer

30

cells. Perhaps we will be able to design compounds that the cancer cells will take up preferentially, thus avoiding the toxicity of drugs.

About twenty years ago scientists discovered that cells infected by viruses produce a compound that interferes with the infection of the cell by other viruses. This compound was named interferon. Later it was found that some compounds resembling nucleic acids can be synthesized in the laboratory, and these can also induce cells to produce interferon. A workshop was organized by the Memorial Sloan-Kettering Cancer Center and the National Cancer Institute to systematically explore the potential of interferon in the treatment of cancer.

I would like to tell you about an idea I have with regard to the invasiveness of cancer, the most feared of its properties. The idea is simple and can be tested. Cancer cells ferment sugar more vigorously than normal cells. The product of that process is lactic acid, which is rapidly excreted by the cancer cells. I have proposed that this acid excretion may have something to do with the invasive properties of cancer. The acid may injure the surrounding normal tissue. The cells die, disintegrate, and thereby allow the cancer to expand. To test this idea we have to find a way to stop the cancer cell from fermenting sugar so fast. How can we do this? The first question we must answer is *why* cancer cells ferment glucose so rapidly. It took us fifteen years to find the answer. Now we know that in most of the tumors we studied, the reason was a leaky pump in the membrane. Our cells have potassium inside and sodium outside, and to maintain this situation they have to pump sodium out and potassium in continuously. This pumping requires energy. Our cells have a wonderful economy in the energy metabolism. They use as the major energy currency a compound called adenosine triphosphate (ATP). ATP is used to synthesize proteins, nucleic acids, and fats; it is used for muscular contraction and for the transport of many compounds, including ions such as sodium and potassium. The production of ATP is very closely coupled to its utilization. Stated otherwise, we only burn as much sugar as we need to generate ATP. I often wished that American industry would take this as a model, rather than produce and then advertise.

What happens when one of the energy utilizing processes becomes defective? What happens when a pump becomes leaky? More energy is needed for the same amount of pumping, and more glucose is therefore fermented to make more ATP. Now that we know the answer to our first question, what do we do with it? We have to isolate the pump and study its properties. We are doing this. We have to find out why the normal cells pump efficiently and why the cancer cell has a leaky pump and hence makes more acid. We are exploring this. We have to search for compounds that will fix the leak of the pump. We are searching for this, and we were lucky—or at least partly lucky. We found compounds that will fix the leak in the cancer cell. Where did we find them? In plant cells! I hope the former director of the American Cancer Society will read this. But I said we were partly lucky because these plant products fix the leak only in the isolated cancer cells. Our blood contains a protein called serum albumin, which prevents these compounds from entering the cancer cell. Now we are trying to find some chemical derivatives of these plant products that will not react with serum albumin. We have some clues and Dr. J. Chopin, a chemist in France, has synthesized a compound which shows some promise in test tube tests. We don't know whether this compound will work in living animals to stop the leak of the cancer cell pump. We don't know whether it will influence the invasive properties of the cancer, even if it fixes the leak and blocks the acid production of the cancer. But we shall try to find out.

Dr. Leon Heppel at Cornell University has made a fascinating discovery. He found that cancer cells, but not normal cells, become leaky when treated with ATP. He and his colleagues are now trying to find the reason for this remarkable difference. But they are also thinking of possible practical applications. Could they take advantage of the fact that in the presence of ATP certain chemicals that cannot enter normal cells do get inside cancer cells? If such a chemical is toxic—for example, for the synthesis of proteins or nucleic acids—would it kill cancer cells and not normal cells? Test tube studies clearly show that this is the case. But will it work in animals? Work on this problem is now in progress.

With these stories I would like to illustrate, once again, how important it is to keep the ball moving. Perhaps our ideas on the effect of lactic acid on the invasiveness of tumors is wrong. However, if they have led us to an analysis of the properties of tumor membranes, the knowledge we gain may help to solve the problem of why tumor cells become leaky when exposed to ATP. But we need to know more about the general properties of membranes of normal cells. We need to know which chemicals can get into cells and which ones are excluded before we can hope to design a series of toxic compounds that will selectively enter tumor cells.

I realize that I have not given a simple formula for the cure of cancer. There is none. I hope, however, that I have established that rational approaches to the prevention and cure of cancer are being vigorously pursued in many laboratories, that because of our limited knowledge we cannot predict which of the approaches is most promising, but that by broadening our basic knowledge we can widen our attack and increase our chances of success.

Science and Diseases of Organs

Heart Diseases and Arteriosclerosis

In 1965 a report was published by the Commission on Heart Diseases, Cancer, and Stroke that had been appointed by President Lyndon B. Johnson. I was a member of one of the panels. I did not get a headache at these meetings but I probably should have. The recommendations of our panel were virtually ignored in the final report, which concentrated on the immediate practical problems of these diseases. Clearly, that was what President Johnson wanted.

A task force of the National Heart and Lung Institute on arteriosclerosis, organized by Dr. T. Cooper, published an excellent report in 1971. Two mottoes preceded the report, citations from the writings of Louis Pasteur:

> To him who devotes his life to science, nothing can give more happiness than increasing the number of discoveries, but his cup of joy is full when the results of his studies immediately find practical application.

> There are not two sciences. There is only one science and the application of science and these two activities are linked as the fruit is to the tree.

Other reports on coronary heart diseases followed, one from the National Heart Foundation of New Zealand and one in 1974 from an Advisory Panel of the British government. In essence, all the reports agree. Diseases of the

heart and blood vessels have multiple contributory causes
and there are several "risk factors." The most important
are hypertension, high lipid levels in serum, cigarette
smoking, and diabetes. But some of the "facts" about
these risk factors are still soft. For example, it seems
clear that treatment of hypertension decreases the incidence of stroke, but the guidelines as to what level the
blood pressure should be lowered to are rather vague. Ac-
curate estimates on the number of cigarettes that cause
damage are also not available. The role of cholesterol and
other lipids in the diet is still controversial. It will become
apparent from the discussion that follows why it is so difficult to obtain unambiguous information on these critical
points.

Arteriosclerosis, a disease of the blood vessels giving
rise to heart attacks and strokes, is the major cause of
death in the United States. In Japan the incidence of
coronary heart disease is about one sixth of that in the
United States. We don't know why. Nutritionists have
blamed the rich American diet, but convincing data are
not available. I mentioned high lipid content (cholesterol
and triglycerides) among the risk factors. This is what
was written in the task force report:

> Elevation of serum lipids is implicated in the etiology
> of arteriosclerotic disease. Current data indicate that
> the average North American has higher than opti-
> mal blood lipid levels and ingests excessive calories,
> saturated fats and cholesterol.

All these reports agree that the evidence for dietary influences is far from convincing. For example, there is good
evidence that in some people an increase in the ratio of
polyunsaturated fatty acids to saturated fatty acids in the
diet leads to a lowering of the serum cholesterol level. Yet
in 1974 the British panel of experts agreed unanimously
that there was no evidence that such a diet regimen had
any influence on the incidence of coronary disease. In

spite of this, to be on the safe side, the New Zealand and the British reports recommended a reduction of saturated fat in the diet. The American task force was more conservative:

> Intuitively, it would seem prudent to decrease the incidence of hyperlipidemia (excess fat substances in the blood) in the population of the United States by controlling diet. However, this would be a formidable venture if it were to involve changing the diet of the entire nation. Indeed before advocating such a major revolution in diet, the Task Force concluded that convincing evidence should be sought that lowering the levels of lipids in blood reduces morbidity and mortality from arteriosclerosis.

Years have passed since the task force report was published, and the role of diet in arteriosclerosis and heart disease is as controversial as ever. It is important that lay people understand why it is so difficult to obtain clear-cut information from clinical investigations in general and particularly when dietary problems are involved. Obviously, we cannot compare studies on Japanese and Americans. There are too many differences—genetic and climatic factors as well as varying dietary, working, resting, and sleeping habits. But actually there are as many differences among Americans, black, red, and white, of different racial origins. The climates we live in and the lives we lead are as different as those between the average Japanese and the average American, whatever "average" means. These variables make clinical research very difficult and require studies with huge numbers of patients.

Compare this situation with the possibilities of laboratory experiments. Animals from pure genetic lines can be used. They are always available for analysis, and they never refuse a post-mortem autopsy. The conditions of feeding, lighting, and temperature can be rigidly controlled. Most important, in some animals diseases such as arteriosclerosis can be artificially induced by an appropriate diet high in cholesterol and other fats. There is however a serious drawback to such animal studies. We often don't know whether the artificially induced diseases are representative of the human diseases. In some instances the similarities are so striking that we do not hesitate to use these animals as models for studies and treatment. In other instances there are considerable doubts. I believe that the similarity of arteriosclerosis produced in animals by feeding large quantities of cholesterol and fats is close enough to the human disease to justify the study of the effect of risk factors such as hypertension and diabetes. These model diseases are particularly valuable for initial trials of therapeutic agents.

Smoking is a Risk Factor

Smoking is a risk factor that can be clinically evaluated more easily than the high lipid diet. There was agreement in all the above-mentioned reports that cessation of smoking decreases the likelihood of death and disability from heart and blood vessel diseases. Smoking is considered the most important risk factor for men who have a coronary attack before the age of fifty. The risk of strokes is also higher among smokers. For all vascular diseases the risks are particularly high among those smoking two or more packages of cigarettes daily from an early age. It takes about ten years of non-smoking to lower the risk to that for the non-smoker. These are data obtained in the United States. In Japan, which has a much lower incidence of coronary heart disease, cigarette smoking does not appear to have any influence.

What can be done about smoking as a risk factor? The task force report made four recommendations: an expanded program of advice to the public regarding the relationship between smoking and heart disease; support of behavioral research aimed at developing more effective methods for inducing people to stop smoking; development of effective programs for the prevention of smoking, particularly in the young; and research into the basic mechanism by which cigarette smoking exerts its harmful effects.

How much has been achieved in the past five years, particularly with respect to the third recommendation about smoking in the young, which is most pressing and immediately applicable? While smoking in the older age group, particularly among the well-to-do population, has declined during the past few years, it has increased among the young. Moreover, once the young ones become "hooked," it is more difficult for them to stop smoking, since unlike the middle-aged patient they are not as worried about their health.

Here is one problem where the basic knowledge (the relationship between lung cancer and smoking) *is* available. Yet we have not succeeded in getting the message to the younger generation in an effective way. They know there is a relation between lung diseases and smoking but do not *really* know it. My daughter smoked when she was in her early twenties. Neither of her parents smoked, and we knew better than to pressure her to stop. One day she visited a friend in the hospital. In the next bed was a man with advanced emphysema, and his misery came through to her. She walked out of the hospital and stopped smoking.

We need to know more about the psychology of our young, about the image they have of themselves, and how we can help to improve it. Good basic and applied research is needed here, ranging from studies in the field of animal behavior and conditioned reflexes to psychological problems in communication between parents and their offspring.

There is another important problem in relation to smoking that needs more study—the general problem of addiction. We need to know more about the biochemistry as well as the psychology of addiction. We need a vigorous and imaginative program, not unlike the one we have against infectious diseases, but at the mental level. There is a greater resemblance between smoking and infectious diseases than meets the eye. The incidence of teenage smoking is much higher among children of smokers than among those of non-smokers. It is well known how an infection of smoking can spread through a school building. Our educational programs against smoking are "too little and too late." Much more should be done, starting at the elementary school level. But we need innovators, people with vision and money.

I have never smoked and probably cannot fully appreciate the difficulties people have when they try to stop. But watching the agony of many of my friends who *have* tried, I am convinced that many smokers are physiologically addicted to nicotine. Thus the problem of smoking could be used as a valuable model for behavioral studies of addiction and may help in the understanding and treatment of alcohol and drug addicts.

Other Risk Factors

Two risk factors emphasized more in the British report than in the previous ones are obesity and softness of water. It is clear from studies conducted by insurance companies that very obese persons are more likely to die of complications of arteriosclerosis than those of normal weight. But since obese persons also have a higher incidence of other risk factors, such as hypertension, high serum lipids, and diabetes, it is difficult to separate any direct effect of obesity. In any case, recommendations in all three reports include elimination of large weight excesses. As many of us know, this is easier said than done because psychological as well as physiological factors play important roles in obesity. Among the physiological factors is the control of energy metabolism which has not been sufficiently studied, as I shall discuss later. On the other hand, much progress has been made in recent years at the fundamental level of obesity because of the availability of genetically obese strains of mice.

The relationship between the softness of water and the increased incidence of heart disease needs more epidemiological studies as well as laboratory experimentation. If soft water can be firmly established as an important risk factor, this should be one of the easiest problems to solve.

What Has Basic Research Contributed?

What about basic research on arteriosclerosis? What can we learn from such studies? One of the most striking facts we learn from the chemistry of the arteriosclerosis lesion is that it is complex. There is proliferation of smooth muscle cells, deposition of lipids as well as of fibrous proteins, and eventually calcification. Little wonder that with so many participating components there are multiple risk factors and a complex clinical picture.

But rather than predicting what basic research will teach us in the future about arteriosclerosis, I would like to point out how much our present methods of diagnosis and epidemiological studies depended on rather unrelated basic science research performed in the past. A chapter in the task force report tells the impressive tale of seven Nobel Prize laureates who have worked on fundamental problems in physics and chemistry. The Svedberg, a Swedish scientist, developed the analytical ultracentrifuge while working on the sedimentation of colloidal metal particles. This instrument later permitted the separation of lipid protein complexes from the blood of patients with arteriosclerosis and other diseases with abnormalities in blood components. Arne Tiselius, another Swede, used an electric field to separate protein components. His methods have subsequently been expanded into clinically applicable procedures used all over the world for the diagnosis of diseases. Similarly, studies of wool proteins by Richard Synge and Archer Martin in England have led to the development of a method called "partition chromatography" now being used for the separation of proteins, lipids, carbohydrates, vitamins, and hormones. Further developments of this method, called thin layer chromatography and gas liquid chromatography, have permitted analysis of lipids in blood and tissues in less than 1/100th of the time required previously. These methods are now widely used in research on arteriosclerosis.

Studies by George de Hevesy in England on the separation of lead and radium led him to the use of radioactive isotopes in biochemistry, a development that has revolutionized biochemistry. These tools specifically permitted Konrad Bloch in the United States and Feodor Lynen in Germany to carry out their important studies on the biosynthesis of cholesterol, which have greatly aided the analysis of the role of this lipid in arteriosclerosis.

Diabetes

One of our senators is particularly anxious to get a quick cure for this disease. Specifically, he wants us to develop an artificial pancreas which will automatically deliver the needed amounts of insulin to the diabetic patient. Do we know enough about diabetes to warrant such an expensive program? Could such an artificial organ be made safe, as safe as a pacemaker for the heart? Can we rely on a computerized organ for millions of diabetics to deliver just the right amounts of insulin needed? This problem is much more complex than that of a pacemaker because the need of the patients varies greatly with their exercise, with the kind of food they digest, and many other factors.

A controlled delivery of insulin is not an impossible goal, but it will require also a monitoring device that will sense the glucose level in the patient.

A more realistic approach was taken by the organizers of an international symposium on diabetes in Jerusalem, under the slogan "A full life despite diabetes." Indeed, diabetes is a disease that can usually be treated with remarkable success in spite of our insufficient knowledge. Appropriately, this symposium was held in 1971, fifty years after the first clinically safe preparation of insulin from pancreas was obtained in 1921. It is clear from this symposium and many others that have been held subsequently that despite the passage of half a century since the discovery of insulin, we still do not understand its mode of action. And we still do not know the nature of diabetes in the majority of human cases. Contrary to popular belief, most diabetic patients do not have an insulin deficiency. Only about 10 percent, mostly comprising the juvenile diabetics, have a pancreatic insufficiency for insulin production. The others suffer from a greater need for insulin or an insulin resistance. Even the excessive amounts of insulin produced by the pancreas of many obese diabetics are not sufficient to keep the blood sugar at its normal level. Recent work at the National Institutes of Health and at Stanford University has shown that there is an inverse relationship between the amount of insulin in the blood and the number of molecules on the surface of cell membranes that interact with insulin. These molecules are called insulin receptors. It appears now that diabetes may be a disease of the membranes; the insulin resistance may be caused by an altered response of the membrane and of the insulin receptors. It has been reported that oral antidiabetic agents, such as the sulfonylureas, increase the number of available receptors on the membrane of diabetic cells.

Why do I mention all this after I supported "a full life despite diabetes," implying that diabetic patients can maintain quite successfully an active life, such as heroically demonstrated by the ice-hockey player Bobby Clark, of the Philadelphia Flyers?

Actually I support the attempts of the senator who would like to free the one million diabetics in the United States from the chains of constant medical supervision. But I urge him not to support an automatic delivery device without a glucose monitoring system. I hope he will read these letters because I would like to convince him that the safest and most rational way to go about this is to learn more about the insulin receptors and the cell membranes, about the mode of action of insulin and of the oral antidiabetic drugs, and to try to induce experimentally adult-type diabetes in animals so that we can study the disease under controlled conditions.

Lung Diseases

In September 1972, former President Nixon signed the National Heart, Blood Vessel, Lung and Blood Act. It requested a coordinated plan from the National Institutes of Health to combat the diseases of these organs. In response, the National Heart and Lung Institute organized twenty-eight task groups consisting of approximately 250 medical advisors, among them professors of medicine from leading universities. The section on priority recommendations on lung diseases reads as follows:

The major missing element in those diseases which comprise the bulk of pulmonary disease is knowledge of the cellular biology of lung tissue. Understanding of this would lead to major advances in chronic bronchitis, emphysema, pulmonary fibrosis, immunologic lung diseases, respiratory diseases of children and respiratory failure. Hence, studies in this area are the number one priority. More rapid progress in this area would be made by the establishment of investigative units organized to explore these basic considerations.

It should be recognized, however, that progress will be slow in this field and that a number of palliative but specific approaches can be implemented at this time. In the case of chronic bronchitis and emphysema, programs designed to decrease smoking, particularly in the young, would make a major impact. In the case of pulmonary fibrosis, the identification of causes would lead to specific measures of prevention. In the case of allergic airways disease, studies of how individuals differ in immunologic pattern and pharmacologic response are recommended. In the case of respiratory failure, the evaluation of existing therapy and dissemination of information concerning diagnosis and treatment are recommended.

One of the reasons why, in the words of the report, "virtually nothing is known of the manner in which pathogenic processes produce destruction and alteration of the architecture of the lung," is that the basic building components of the lung are very complex molecules. The lung is particularly rich in two rather unusual proteins—collagen and elastin—which are essential for the tensile strength and collapsibility of the lung air sacs (alveoli) during respiration. Because of their peculiar properties, these proteins are very difficult to analyze, but in recent years good progress has been made in our understanding of their chemistry and biosynthesis. But little is still known as to how these proteins are degraded within the lung, yet this knowledge may be a key to the problem of the disabling lung disease called emphysema. In this disease the thin walls of the air sacs lose their elasticity and then tear.

Emphysema is among the leading causes of disability and death in the United States. Moreover, the death rate has almost tripled in the past twenty years. There are

44

presently about one million emphysema and four million chronic bronchitis patients in this country.

Can we prevent and cure emphysema? From what we know about the chemistry of the lung proteins, it seems highly unlikely that we will be able to cure emphysema. The damage appears to be irreversible. But it should be a very simple matter to prevent emphysema because the most important factor in this disease is smoking and pollution. The task force report states: "If it were possible to stop cigarette smoking the problem of chronic bronchitis and emphysema would largely disappear."

Recently, while I was composing this letter, the American Cancer Society launched an educational program against smoking. Representatives of the tobacco industry were indignant and told the American Cancer Society to stick to basic research instead of getting involved in propaganda. It is perfectly all right to seduce our younger generation into smoking by romantic advertisements, but to give them the hard facts about smoking is labeled propaganda! Among the information distributed by the Society was the fact that the United States government donates $60 to $80 million annually (over $4 billion since 1937) to tobacco growers.

Basic Studies on Protein-Degrading Enzymes

Not all smokers get emphysema. Recent research has shown that there is also a genetic risk factor. Individuals who inherit a deficiency in a blood component called α_1-antitrypsin are predisposed to develop emphysema. Trypsin is a protein-degrading enzyme and antitrypsins are compounds that inhibit this enzyme. Could we use a fraction of the 80 million dollars given to tobacco growers to support instead an inexpensive study of the relationship between α_1-antitrypsin and emphysema and to develop a routine test that could be performed in our genetic centers?

I mentioned earlier that the physical damage in emphysema is irreversible. We therefore badly need methods for the early diagnosis of this disease before it is too late. Animal models are now available. It is possible to inject into the lungs of rabbits an enzyme called papain which in a few days produces emphysema. Papain is like trypsin, an enzyme that degrades proteins. You can see now how the genetic deficiency of α_1-antitrypsin may be related to the susceptibility to emphysema. I would like to see some of the eighty million dollars given to tobacco growers spent on the effect of trypsin on lung proteins, and on the effect of protein-degrading enzymes of white blood cells (that have been implicated in the disease process) on elastin and on other proteins.

Further research on α_1-antitrypsin is important beyond its significance for lung diseases. A disease (cirrhosis in infancy) involving damage to the liver has been correlated with α_1-antitrypsin deficiency.

I do not want to give the impression that these basic studies are simple and could give us quick answers. I want to take this opportunity to illustrate how complex nature is, even when we are dealing with an apparently simple protein molecule such as α_1-antitrypsin. There are at least twenty-one genetic types of α_1-antitrypsin and nothing is known about their biochemical differences except that they can be separated in an electric field. Moreover, human blood contains, in addition to α_1-antitrypsin, at least five other inhibitors of protein-degrading enzymes.

Eighty million dollars spent on research of inhibitors of protein-degrading enzymes could influence our lives far beyond the prevention of emphysema. Such inhibitory proteins are present in semen and have to be removed before fertilization can take place. Inhibitors of protein degradation have been put into vaginal jelly and were found to be effective contraceptives in rabbits.

Scientists have proposed and obtained some evidence

that protein-degrading enzymes play a role in the growth and invasive properties of cancer. Many laboratories are presently studying the effect of inhibitors of protein-degrading enzymes on the growth of tumor cells.

Inhibitors of protein-degrading enzymes play an important role in the life of a parasite called *Ascaris lumbricoides*, a roundworm resembling the earthworm. It is estimated that one quarter of the world's population is infected with human *Ascaris*; the infectious parasite is even more popular among hogs. There are some interesting differences in properties between the inhibitors of protein-degrading enzymes of the human and hog parasites which correlate with the fact that the hog parasite cannot infect humans and the human parasite cannot infect hogs.

Protein-degrading enzymes play an important role in blood coagulation, which can be influenced by inhibitors. Much more basic work needs to be done in each of these areas.

Diseases of the Kidney

Among the diseases of the kidney I have selected the nephrotic syndrome for discussion. It gives me an opportunity to make some comments about the importance of immunology research and about the formation of antibodies against our own cell components, leading to so-called autoimmune diseases.

Nephrosis is not really a disease by itself because it has multiple causes and sometimes arises as a complication of other diseases, such as diabetes. But it is clear that in a large proportion of cases the disease has an immunological origin.

The disease can be induced in laboratory animals, and although there are differences from the human syndrome, we have learned a great deal from experimental nephrosis. Antibodies produced in a rabbit against the proteins of rat kidney cause a nephrotic syndrome when injected into rats. It was shown that the antibodies formed against the basement membrane of the filtration apparatus of the kidney were responsible for evoking the disease. Considerably more information is now available about the changes in the membrane caused by the immunological interaction between the protein and its antibody. Of even greater interest is a strain of mice which when crossed with another strain gives rise to an offspring that is uniquely susceptible to a kidney ailment which has the characteristics of an autoimmune disease. This is a disease in which the body produces antibodies against some of its own proteins leading to damage of some organs. These mice serve as good models for the treatment of autoimmune diseases with so-called immunosuppressive drugs. Immunosuppressive drugs are now clinically used in many connective tissue diseases and in organ transplantations. I mentioned earlier the difficulties of clinical research in general. The evaluation of drugs in human diseases that are often of different origin is not an easy matter. Obviously, in the last analysis treatment of human patients is the acid test, but appropriate animal model diseases are of enormous value in the early stages of evaluation.

Great strides have been made in research in the area of immunology and immunogenetics. But as with all other fields of science, rapid progress is always associated with the opening up of new basic concepts that need to be explored.

Patients with chronic kidney diseases, under treatment with hemodialysis (a form of artificial kidney), or patients with kidney transplants have a high incidence of a bone disease called renal osteodystrophy. Many conferences have been held in the past ten years on this subject including one under the auspices of the Artificial Kidney-Chronic Uremia Program of the National Institutes of Health. In my last letter I mentioned monovalent ions such as sodium and potassium, atoms that carry a single positive charge. In osteodystrophic patients there is a disordered metabolism of divalent ions (carrying two positive charges), particularly calcium, an essential component of bone. Vitamin D is an important regulator of calcium metabolism. I am sure you are familiar with the classical story of rickets and the role that basic research has played in the prevention of this crippling disease. Patients with renal osteodystrophy become vitamin D-resistant—the vitamin no longer acts properly. About nine years ago it was discovered that vitamin D is converted in our body to another compound which is the physiologically active agent. The enzyme involved in this conversion is in the kidney. When the kidney is damaged and functions improperly, the active compound is not formed and calcium metabolism becomes deranged, leading to the bone disease. The story is more complex than I have described, involving regulation by a hormone and other factors. But the main point is that these basic studies have induced chemists to synthesize in the laboratory the new vitamin D derivative which is now being used successfully in the treatment of renal osteodystrophy.

In this letter I have selected for discussion diseases such as emphysema, nephrosis, and diseases of the heart that render patients chronically ill. There is much suffering associated with these diseases—pain, discomfort, both physical and psychological; and the financial burden of hospital bills and of treatment at home is enormous. Of course we want to prevent these diseases and treat them once they have inflicted themselves upon the patient. I have suggested plans that could be set in action soon and plans that must await the research of tomorrow. But while we advocate more research we must be aware of what we are doing to our population. We have steadily increased life expectancy, but have we properly looked at the other side of the coin? What has our society done to make people want to live longer? It seems that for some, it is better to smoke and die coughing miserably as an emphysema patient than to remain well and become old and live a life of loneliness and despair. Many of the Asian countries do better for their elderly than we do. Once again we need individuals with vision who will build into our society a place for the old. It has been said that now that we have learned how to put more years into our lives, we have to learn how to put more life into our years.

Science and Genetic Diseases

There is a widespread attitude among physicians and lay people that genetic diseases are hopeless and that there is little one can do. Well, that's not true.

Let me start at the beginning. I tried to illustrate in my previous letters how all diseases in a sense are genetic. Schizophrenia has been called a polygenic disease (perhaps multifactorial is a better term) because several conditions appear to be required for the appearance of the manifestations of the disease. I have elaborated on a similar multifactorial complexity in cancer and in diseases of the organs and membranes. Even infections can be shown to be influenced by genes. Some cells that are susceptible to a virus or to a parasite can be made resistant by a single mutation, i.e., by the alteration of a single gene.

Inborn Errors of Metabolism

But what about the *bona fide* genetic diseases where we are faced not only with a change in susceptibility to disease, but with a direct causal relationship, such as the lack of an essential enzyme catalyst in the body? There are hundreds of such genetic diseases. In some cases we know the exact biochemical lesion, as in phenylketonuria (PKU), galactosemia, Pompe disease, Lesch-Nyhan disease, and many others. Some of this knowledge was made possible by recent developments in tissue-culture techniques that make it possible to culture in the laboratory the diseased cells of the patients. Hundreds of additional genetic diseases have yet to be identified in this manner. Will this understanding help us to prevent and treat some of these diseases? The answer is yes, because we already have an impressive record of achievement.

PKU, which I mentioned in my first letter, is not as rare as some people think, since it affects about one out of ten thousand newborn babies. It is a depressing experience to watch these mentally defective children in an institution. We know exactly why they are sick. They are missing a single protein catalyst, an enzyme called phenylalanine hydroxylase, which metabolizes the amino acid phenylalanine. If this amino acid is not properly metabolized, the blood contains high levels of phenylalanine which, together with toxic side products that are formed, damage the developing brain. But if we delete phenylalanine from the diet at an early age, much of the damage can be prevented. The adult brain is less sensitive, and alternative routes of phenylalanine utilization have been reported to develop in later life. In any case, the rigid adherence to a phenylalanine-free diet, which is expensive and difficult to maintain, can be relaxed after the brain is developed. It should not be hard to convince anyone that we needed the fundamental information on the metabolism of the amino acid and of the enyzme in order to discover the defect and to design the proper diet for these unfortunate children.

It is appropriate to point out that problems of this type are not solved without paying a price. Errors in diagnosis have been made; not all babies with high phenylalanine in the blood suffer from PKU. The diet given to PKU patients is low in protein and probably not optimal. It is very unpalatable and needs improvement, both nutritionally and gastronomically.

Another genetic disease susceptible to treatment is hereditary galactosemia. Galactose is a sugar present in milk. Children with this disease do not tolerate milk and become very sick and may die when they drink or eat food containing galactose. If galactose is rigorously avoided, the disease can be suppressed. Galactosemia, PKU, and other treatable diseases can be diagnosed at birth.

But, if possible, prevention should start before birth. Genetic counseling of prospective parents is not practiced sufficiently in the United States. Most of our physicians have very poor training in genetics, and indeed it would be difficult for practitioners to acquire expertise in such a complex and rapidly expanding field. But what they should know—and many don't—is that there are medical genetic centers such as the National Foundation-March of Dimes, and the National Genetics Foundation, which can direct physicians and patients to centers in their vicinity for counseling and testing. Patients know about familiar diseases like PKU, cystic fibrosis, Tay-Sachs, or muscular dystrophy. Many parents know about the risks of such diseased offspring and its implications. But they should learn more about the different susceptibility of various populations to Down's syndrome (mongolism) or dwarfism where the risk can be very low or may be one out of four, depending on the type of genetic defect. Proper diagnosis of the genetic background of each of the parents and counseling can be obtained from the genetic centers. The final decision about child bearing is and should be with the prospective parents. However, they should be clearly aware of what the risks are, and they should know what it is like to have a defective child. A woman who has been saved from PKU mental defectiveness should be told that a child of hers will be exposed to the high levels of phenylalanine in her blood and will be mentally damaged. I spoke about education in my last letter. This is another area where more needs to be done.

We can now move to the second stage of disease prevention. The first genetic disease treated prenatally was a disease of the blood called erythroblastosis fetalis. It occurs when the child's blood type (Rh) is incompatible with that of the mother, who forms antibodies against it. Blood transfusions into the fetus *in utero* can be performed. Better still, an Rh-negative mother, previously exposed to an Rh-positive child, can be desensitized before a second pregnancy takes place. This probably represents the first conquest of a genetic disease. Another example of prenatal treatment of a genetic disease was reported recently. The disease is methylmalonic acidemia, which is an inborn error of metabolism affecting the amino acid isoleucine and resulting in high levels of methylmalonic acid in the blood and urine. In some of these patients (but not all), the defect is caused by an abnormality of vitamin B_{12}. If amniocentesis (a surgical penetration into the sac that surrounds the embryo) is performed early, the amniotic cells can be grown in tissue culture and the metabolic lesion identified. If it is the vitamin B_{12}-sensitive form of the disease—as it was in this case—the fetus can be treated by giving the mother massive doses of the vitamin, and the disease can be prevented. Since a mother pregnant with such a diseased child excretes large amounts of methylmalonic acid, the diagnosis of the disease (but not the vitamin B_{12}-sensitivity) can be made by urine analysis. I should like to point out here in passing that most of the basic information on methylmalonic acid metabolism and on vitamin B_{12} has come from studies on bacteria rather than on mammalian cells.

PKU and galactosemia illustrate one principle in the treatment of genetic diseases—dietary avoidance of food substances that are improperly metabolized. In erythroblastosis fetalis, a damaging antibody is being removed. Methylmalonic acidemia illustrates a third principle—replacement treatment with a metabolic product that is not properly synthesized by the patient with the genetic defect. Treatment of juvenile diabetics with insulin is another and more common example. A fourth possibility is illustrated by genetic diseases elicited by drugs. There are some genetic aberrations that cause changes in enzymes, e.g., in red blood cells, that give rise to severe anemias when the patients are exposed to drugs, such as primaquine or aspirin, which are quite safe for normal subjects. The treatment is simple—avoid the drugs. A fifth example for the treatment of a genetic ailment is illustrated by Wilson's disease. In this disease there is an abnormality in copper metabolism resulting in deposits of copper in the liver and the brain. The disease can be treated by administration of copper-binding agents such as penicillamine. Actually, this is one example of the successful treatment of a genetic disease in which we do not have a clear understanding of the biochemical lesion. The contribution from basic research came from chemists who examined the ability of various chemicals to bind copper.

The sixth possibility of treatment of genetic diseases is replacement therapy—incorporation of the missing catalyst into the deficient cells of the patients, or transplantation of organs that will supply the missing enzyme. Kidney transplants have been successfully used in patients with Fabry disease, in which a kidney enzyme is defective. Even liver transplants have been reported with some success in genetic lesions of liver enzymes. Another exciting approach of therapy which has been tried in several genetic diseases during the past twelve years is to give the patient the missing enzyme. Thus far this has been uniformly disappointing even when enzymes of human origin were used to avoid immunological side reactions. But enough information was gained from these attempts to realize some of the difficulties in the approach. At the National Institutes of Health, Dr. Roscoe Brady and his collaborators injected patients suffering from Tay-Sachs disease and Fabry disease with the specific enzymes they lacked. Within twenty minutes, most of the enzymes had disappeared from the bloodstream, and none seemed to have reached the brain where it was needed. The clinical condition of the patients was not improved, although it was clear that the enzymes had acted on their substrate in the blood and had reached the liver. In fact, there was some indication that the active enzymes may have reactivated some of the inactive, aberrant, enzymes in the liver of the patients. But the fact that the enzymes did not reach the brain and were so rapidly eliminated explains these and many similar failures.

What is needed is some method to protect the administered enzyme and to allow it to "home in" on the tissues where it is required. For the enzyme to be effective it must be incorporated into the cell and survive there. Cells can incorporate particulate matter, e.g., an enzyme, by a process called endocytosis, and there are some indications that the efficiency of this process can be influenced by alterations in the chemistry of the ingested enzyme without interfering with its ability to act as a catalyst. These studies are in their infancy, and much more basic research is needed before practical applications should be attempted.

During the past few years another more sophisticated approach to this problem has been taken. Investigators have learned how to incorporate enzymes into little artificial spherical compartments made out of lipids. These tiny particles are called liposomes. It was found that when cells of patients with certain genetic diseases are grown in tissue culture they are capable of incorporating these enzyme-containing liposomes and can thereby be "cured." However, when experiments were performed with live animals, considerable difficulties were encountered. One of the major problems was that liposomes injected into the bloodstream were recognized as "foreign," taken up by the scavenger system of our body—the "reticulo endothelial system" of the liver—and deported via the bile ducts into the intestine. It was therefore necessary to construct the liposomes in such a manner that they operate as guided missiles, with an ability to "home in" on a particular cell or tissue. A good model for such

56

studies was found in the dogfish (a toothless, small shark) which lacks an oxidizing enzyme present in normal human cells as well as in plants. The enzyme (prepared from horseradish!) was put into the liposomes, which were treated with an aggregated immune protein. These liposomes were actively taken up by the white blood cells of the dogfish. This opened up an exciting possibility. Dr. Brady proposed, after his failure with injection of free enzymes, that an enzyme "linked to the patient's own leukocytes" (white blood cells) could help to cure a genetic disease such as Tay-Sachs if the white blood cells could be induced to enter the brain. In the case of the white cells of the dogfish this was shown to be possible. After injection of an irritant, the white cells containing the horseradish enzyme entered the brain! It was also found that human white cells of patients with Tay-Sachs disease could take up liposomes containing the missing enzyme. But ways will have to be found for the white cells to discharge the enzyme in the brain, and some ideas as to how this could be done have been suggested.

Although we are by no means ready to cure patients with these particular genetic diseases of the brain, I believe we are on the right track.

Cystic Fibrosis

I hope I have shown that we have several approaches to the treatment of those genetic diseases which we understand. Unfortunately, there are still many diseases with yet unknown causes. Cystic fibrosis is one of them. The incidence of this disease ranges from one in 2,000 to one in 15,000 in different populations of Caucasian descent; it is considerably lower in non-Caucasian populations. More than 50 percent of the afflicted die before the age of twenty-one. Clinically, the most distressing manifestation (which usually is also the cause of death) is an abnormality of mucus excretions which block passages of the lung, liver, pancreas, and intestine. The sweat of children with cystic fibrosis has a high content of sodium chloride (table salt), an important feature aiding in the diagnosis. However, the primary lesion is unknown.

It is often difficult for a lay person to grasp the difficulties associated with the identification of the primary lesion of a disease. Why has the extensive research supported by the Cystic Fibrosis Foundation failed as yet to come up with an answer to this fundamental question? It seems not unlikely that in cystic fibrosis there is something wrong with the transport of ions, such as sodium and calcium, into and out of cells. Unfortunately, our basic knowledge of these transport processes is still very rudimentary. For example, in the case of sodium ions, only one of the transport mechanisms has been relatively well characterized, namely that of the sodium pump mentioned in my second letter. There are at least three other sodium "channels" which allow the transport of sodium from the inside to the outside of the cell and vice versa. The characterization of these channels has only just begun. Why is it so difficult to study them? In order to understand biological processes we have to remove from the cell the components and catalysts which facilitate these processes, separate them from other components, and study them in the isolated state so that we can learn exactly how they accomplish what they are doing. It was possible over fifty years ago to isolate insulin from pancreas because there was an assay. If insulin is injected into an animal, the blood sugar level falls. This can be measured accurately and can be used to standardize the potency of the insulin preparations. How can we assay a sodium channel? There are two possibilities. First, there are drugs which interact specifically with some ion channels. For example, the puffer fish produces a lethal toxin called tetrodotoxin, which specifically interacts with one of the sodium channels present in the brain. We can use this specific interaction for an assay of this ion channel.

Several research groups are now trying to do this. Second, if we don't have such a "marker" substance we can prepare liposomes, the tiny artificial compartments made of lipids, and try to incorporate a sodium channel into them. Since liposomes are closed compartments, we can measure fluxes of ions into and out of these vesicles.

Perhaps one day we will succeed in isolating from the mucus-secreting gland the sodium channels of normal and cystic fibrosis cells and put them into liposomes. Perhaps after years of study we shall discover that there are differences between them. This will be an exciting step forward. Will it help us with the treatment of cystic fibrosis? Perhaps, by the replacement therapy method described earlier in this letter. But perhaps we shall discover no difference in the sodium channels of normal and diseased cells. What then? Will all our efforts have been wasted—at least as far as cystic fibrosis is concerned? I think not, because we shall have on hand a simple new tool with which we can explore other possibilities as to why sodium transport is defective in cystic fibrosis. Ten years ago investigators discovered that in the sweat of patients with cystic fibrosis there is a "cystic fibrosis factor" which inhibits the uptake of sodium by mucus-secreting glands. Methods using glands, where many complex metabolic events take place simultaneously, are difficult. It would make it so much simpler if the cystic fibrosis factor inhibited sodium transport in reconstituted liposomes. Thus, we would have a simple system to study the mode of action of the factor and reap benefits from the studies with the liposomes which have failed us in the earlier project. We could design a rapid assay for the purification of the cystic fibrosis factor. Progress is slow now because the assay is difficult. Nevertheless, some information on this factor is available; for example, it is positively charged. It has been reported that negatively-charged compounds such as heparin, the well-known anticoagulant of blood clotting, counteract the inhibition of sodium transport by the cystic fibrosis factor. It is not difficult to imagine how such studies might one day lead to a rational treatment of the disease.

Familial Hypercholesterolemia

Patients with this genetic disease show high blood levels of cholesterol and changes in blood vessels. Death from heart attack often occurs before the age of thirty. Recent biochemical studies have contributed greatly to our understanding of the defect, which can be analyzed in skin cells obtained from the patients and grown in the laboratory in tissue cultures. The synthesis of cholesterol is normally controlled by lipoproteins (fat containing proteins) of the blood. It appears that the cell surface membrane of patients suffering from *familial hypercholesterolemia* is defective in its interaction with lipoproteins. Although we do not know as yet how to treat these patients, it is now possible to study the diseased cells in test tubes, and we can look for chemicals that will repair the defective membrane just as we have done in the experiments with defective membranes of cancer cells described in letter 2.

We cannot charter these applications; they have to emerge, as Pasteur said, like the fruit from the tree. But we have to see that the tree of science can grow with sufficient nourishment and free of injuries inflicted upon it by the changing moods of society.

Social and Ethical Problems

A problem often raised in connection with PKU is that by keeping the patients alive we may be harming the gene pool. Are we propagating the disease by allowing these patients to become normal and to reproduce? According to experts in population dynamics, this is not a serious problem with a recessive disease such as PKU. The gene fre-

60

quency for this disease is 0.008, which can be expected in one hundred generations (2,000 to 3,000 years) to rise only to 0.0096 if every affected child born in the United States is treated and reproduces. For dominant diseases, the effect on the gene pool would of course be much greater.

Another problem with PKU is educational. Adult females cured by the diet treatment who become pregnant should go back to a low phenylalanine diet. I have mentioned earlier that the adult brain is much less sensitive to high doses of this amino acid than the growing fetus. Thus, even a normal child may be harmed in the uterus when exposed to the high levels of phenylalanine in the blood of his mother. It took a while to learn this lesson, and it requires some sacrifice on the part of the mother to go back to the rather unappealing diet of the first years of her life.

The economics of testing for genetic diseases deserve discussion. For example, in Massachusetts one of the most active genetic screening programs was initiated in 1962 starting with the PKU test. Virtually all babies were tested for several genetic disorders. Seven years later an estimate revealed an impressive saving for the state. At that time the cost of the program was $175,000 per year (less than $2 per child), and in one year thirty children with genetic diseases were identified. Eight had PKU, were treated, and escaped mental retardation. Actually, four additional children had high phenylalanine levels, but did not have PKU and did not need the diet treatment. The cost of institutionalizing eight PKU patients for life was estimated in 1970 to be about $1.5 million.

What about abortions? Today the technique of amniocentesis allows us to diagnose many, though by no means all, genetic defects. For example, we cannot recognize congenital heart diseases, other malformations, and many forms of mental retardation. But it does allow us to

SCIENCE AND GENETIC DISEASES

diagnose Down's syndrome (mongolism), a disease due to the presence of an extra (No. 21) chromosome. The risk of this happening is small in young females, but is about 2 percent in pregnant women beyond age forty. The risk is even greater in young females who have had a previous child with Down's syndrome. Some one hundred different recessive genetic disorders can now be diagnosed by growing amniotic fluid cells in tissue cultures and determining the presence of an enzyme defect. These include the dreaded Tay-Sachs disease, Gaucher's disease, and many others. Congenital heart diseases and neural tube defects are the most common birth defects. The latter can be diagnosed prenatally; the former cannot. Progress has been made in recent years in the prenatal diagnosis of sickle cell anemia and thalassemia, diseases of the blood. Several sex-linked diseases, such as hemophilia, can be recognized. Male fetuses have a 50 percent chance of having the disease. Since sex determinations can now be performed accurately on the cells obtained by amniocentesis, the parents can make a decision whether they wish to take the risk or try another time.

The procedure of amniocentesis is safe. A study sponsored by the National Institutes of Health was carried out a number of years ago on over one thousand cases. There was no increase in the frequency of spontaneous abortions compared to a control group (it actually happened to be somewhat less). In three cases there was an error in the prediction of sex presumably because the mother's cells rather than the infant's cells had been cultured. Recent improvements in the technique reduce errors of this type even further.

In my opinion amniocentesis is greatly underused. I recommend it to every pregnant woman beyond the age of thirty-five and consider it mandatory when there is a family history of Down's disease or certain other genetic disorders. Dr. John Littlefield of Johns Hopkins Medical School has pointed out that from a purely economic point of view it would actually be sensible for the government to pay for the amniocenteses in pregnant women aged thirty-five and older.

Amniocentesis is used to collect information for diagnosis and therapy. If the parents decide, for religious or other reasons, to proceed with the pregnancy, they are at least spared the shock of the discovery at birth (or later) of having a diseased child. The decision is theirs.

How many women who have a defective child give birth to additional children? How many women who have had abortions for genetic reasons become pregnant again? I don't know whether such a study has been undertaken, but I venture to guess that a woman who had an abortion for genetic reasons is more likely to want and get another child than a woman burdened with the care of a defective child. This is an interesting question to be considered by the church.

The field of genetic diseases, perhaps more than any other, has depended on basic research. The isolation of human cells and their successful growth in "tissue cultures" was a major breakthrough that required patient research of many years. Only now are we beginning to learn about the vitamins and growth factors that are required for the growth of different cells, and a new field of nutritional and cellular physiology is in the making. For example, certain types of nerve cells have been successfully grown in tissue cultures, but in earlier experiments they did not resemble nerve cells either in appearance or physiology. But now a specific factor has been found that when added permits the growth of cells that *do* behave like nerve cells.

Each genetic defect is a challenge for biochemists. It often takes years of laborious research to identify the enzyme catalyst that is missing or defective. I hope I have succeeded in illustrating that without this basic knowledge it is difficult to design a rational treatment for such diseases.

63

Letter 5

Society and Science—
Funds for Basic Research

Why Basic Science Research?

In 1831, Faraday conducted experiments on electromagnetic conduction which formed the basis for the electric motor. When he publicly demonstrated his experiments, Gladstone (later Chancellor of the Exchequer) asked him about the practical value of electricity: "One day, Sir," Faraday replied, "you may tax it." May, indeed!

Fortunately, there were always men of vision who supported basic sciences. In 1900, E. W. Rice, Jr., the technical director of General Electric (and later its president), created the first basic science laboratory attached to industry in the United States. This famous laboratory housed many brilliant men including Irving Langmuir, a chemist who worked there for forty years and won the Nobel Prize in 1932 for his work on surface films. Many companies followed suit and now virtually all large companies have basic research laboratories.

I remember an interesting luncheon discussion with a group of research directors at DuPont in Wilmington, Delaware, in 1961. I needled them a little by asking them why they were spending so much money on basic research which would not bring them any profit. For example, they have a research group doing outstanding work on nitrogen fixation in bacteria. I asked whether they hoped that bacteria would compete with the industrial yearly production of 20 million tons of cheap ammonia. No, they admitted, they would never make ammonia with bacteria, but they also said that no one could predict what other benefits might come from this basic research work. They admitted that they had been working on nitrogen fixation for many years without profit, but they were not disturbed by this. First, they remembered that the discovery of nylon and plastics at DuPont started with fundamental work in an area of polymer chemistry which at that time seemed as far removed from profit-making as biological nitrogen fixation seems today. Yet with sales of over seventy million pounds per year, nylon turned out to be one of DuPont's major money-makers. In essence, the directors said, "You can't really tell; you have to take chances."

The second reason they gave me for support of fundamental research in Wilmington was even more interesting. They said that DuPont could attract first-rate scientists because the company was known to support basic research. Having these outstanding men rub shoulders at lunch, cocktails, and dinner parties with investigators in the applied field paid off in terms of ideas transmitted and advice given. Langmuir is a case in point, since while working on surface films he made notable contributions to the development of vacuum tubes which were sold by General Electric.

What is good for General Electric and DuPont may also be good for the country!

American Society and Science

During the past fifty years, the relationship between American society and science has undergone dramatic changes, somewhat similar to those of a young man's awakening interest in women. In boyhood he couldn't care less; in adolescence the female enters his life in a storm and all kinds of things can happen. The relationship may end in disaster, in a beautiful union of love, in indifference, or often in a mixture of all three.

Before World War II, American society couldn't have cared less about science. At best, scientists were subjects for cartoons and condescending affection. They were at the fringe of society and had little influence. After the atomic bomb and particularly after Sputnik, the relationship changed dramatically. This is illustrated by figures for federal support of research and development. It increased from $74 million in 1940 to $16.7 billion in 1967. During that period basic research led to revolutionary discoveries ranging from transistors, lasers, and automation to antibiotics and polio vaccine. As the spinning machines and steam engine in the eighteenth century started the Industrial Revolution, the computers are bringing about an equally startling revolution in the world of white-collar workers.

Between 1945 and 1968, "science" was spelled with a large S and idolized in an adolescent type of relationship. The younger generation, a sensitive barometer of the state of society, flocked to the science departments of universities. To become a respected scientist was a status symbol of that society, reflected in the rising salaries of university professors.

However, following this period, the relationship between science and society changed rather abruptly. Considering inflation, federal support started to decline instead of going up. By 1974 support for the physical sciences was 25 percent below the 1969 level.

Today, science is under attack. Some call research an expensive luxury; others look at science as the creator of pollution, as an evil, obstructing a free way of life. For the radicals among the younger generation, science is part of the establishment, aiding the military-industrial complex in its course toward war. But even conservative young people seem to have lost the spirit and enthusiasm for science, partly discouraged by the prospect of facing unemployment after completion of a Ph.D. Since pollution is in fact a product of our technocratic life, some scientists themselves have raised the question whether progress in technology without control is in the best interests of the future.

Reassessment of the Role of Science

Congressmen have started to publicly challenge basic science as a productive activity and to ask for quick solutions of acute problems. The sudden change in the relationship between science and society has positive as well as negative consequences.

Among the positive consequences are a greater awareness and sharper focus on some major problems. We now begin to have a more realistic evaluation of the consequences of technological developments. You will probably agree that twenty years ago the development of supersonic transport would have passed Congress with little opposition. A new concept is emerging, that of "technological assessment." Advisory boards are being appointed to the House of Representatives and the Senate to evaluate the future effect of technological developments. We now realize that the sinister aspects of technology, such as pollution, crowded airways, and cities with their sequels of slums and ill health, have emerged because of lack of foresight and research planning on the one hand and inertia of society in protecting itself on the other.

Let us look at what assessment boards could do in the future. Provided they can be set up in isolation from politics and pressure groups (which admittedly is difficult), they could advise government and industry in planning ahead with innovations without introducing undue risk to the environment. But there is much conflict of interest, and safe solutions are often expensive or inconvenient for the producers. This is why assessment boards are not always popular. Admittedly, there is also a negative aspect to assessment boards. They may delay progress by playing safe. We see this now in the drug industry. The obstacles which the Food and Drug Administration is putting in the path of pharmaceutical companies attempting to develop new drugs are so great and the costs so enormous that it does not pay to develop a drug unless it has a potentially large market. I know of one example which I think is most disturbing. In the laboratory of one of our blue-chip companies, a discovery was made which looked like a good prospect for the treatment of a fatal but rare disease. The company decided not to go ahead with the development of the drug because the testing for safety and effectiveness will cost them much more money than they could possibly recover from selling it later. Could we not think of some way out of this state of affairs?

But don't misunderstand me. I am still listing the assessment boards as an important positive feature in our society. I have participated in advisory boards for the U.S. Public Health Service and on presidential advisory committees. There is a great deal of dedication and goodwill. Many advisers get little or no pay. However, we are human and make mistakes. This is true of scientists, of assessment boards, and of members of Congress. We realize that we cannot go on as we have in the past. Until recently, a technological advance was permitted to be exploited by the industry until damage was done. Atmospheric pollution by cars and accumulation of pesticide residues could have been predicted if we had consulted alert advisers. Both could have been prevented by more intelligent use of our knowledge before the damage was done. Now DDT, which was an enormous life-saver, has a black eye because we did not always use it intelligently.

Fluctuations in Funding for Basic Research

Now to some of the negative aspects which have arisen from the change in the relationship between science and society. One of the most damaging is the abruptness with which the government has cut the funds for research and development. Again, we are dealing with a case of over-reaction. Obviously, the exponential growth of support witnessed in the years after the war could not continue indefinitely. I remember when our daughter was a baby, I plotted her weight gains week by week and calculated the date at which she would reach the weight of an elephant if she continued at that rate. However, her weight gain changed in due time and slowly converted to a pace appropriate for an attractive young lady.

Instead of allowing for such a natural decline in the growth rate of support for research, funding was cut abruptly, in some instances bringing severe casualties to research projects. More serious, since they threaten the nation's future, are the cuts that were made affecting the training of students? If we want to continue to make important contributions in science as we have during the past twenty years, we cannot afford to damage our graduate education. At present, universities are turning back good applicants to graduate school because of lack of funds. Government training grants for student support have been suddenly discontinued. Jerome Wiesner, former Science Adviser to President John F. Kennedy, predicted that unless trends are reversed, the United States would be a very sick country technologically 10 years from now. The cost of fluctuations in support of research is enormous. Well-trained personnel lose their jobs because of withdrawal of funds. Then someone discovers a few months later that this was a good project and money is refunded. Usually new personnel have to be hired and retrained. Well-trained physicists with a Ph.D. cannot find jobs and start to change fields. In fact, we have applicants with Ph.D.s in physics who want to study biochemistry because the field is still being supported reasonably well. But for how long?

It is of interest to look at what happened in Germany

when Hitler curtailed basic research and dismissed professors with Jewish backgrounds or anti-Nazi sentiments. The result was that Germany, a recognized leader in science, fell so far back that it took about twenty years to recover. We no longer require our science students to master German. Why? Because the Germans now publish largely in English to make sure that Americans will read their papers.

What can we do to prevent a similar remission in the United States? Obviously, we have to convince the American people that in the interests of our future, basic science should be supported without abrupt changes. We have to convince people that a simple way of avoiding fluctuation is to make a firm commitment, perhaps a fixed percentage of the Gross National Product, to research and development. At present, that percentage is smaller in the United States than in many other countries including Sweden, Germany, and Japan. Whatever the amount of support, it should be consistent with the increased cost of living and growth of the nation, and it should be immune to influences from pressure groups and politics.

How to Allocate Funds: Some of the Wrong Questions

This is the most important aspect of my proposal, and will require new legislation sponsored by congressmen with vision. They will need more specific advice to design it. How much money should be allocated for applied, how much for basic, and how much for clinical research? Actually, as I will elaborate later, the problem of allocation of funds need not be as complex as it is now. But we first have to ask the right questions. The questions raised above are the wrong questions because we simply cannot give logical and honest answers to them. Let me cite the example mentioned in my first letter—the study of the chemistry of methylene blue, which led to the synthesis of phenothiazine drugs. This would have been labeled by a panel of scientists as basic research, although a director of a dye industry would have supported it as an applied problem. The work on the role of histamine in allergic phenomena could have been labeled as either basic or applied research depending on how the work was planned, but certainly could not have been labeled as relevant for psychiatry at that time. The work on the effect of antihistaminics on shock was clearly internal medicine, but what emerged was more important for behavioral science and psychiatry.

Some members of Congress have repeatedly asked the wrong questions. About twenty years ago I served on the microbiology study section of the National Institutes of Health. The name of the institute to which Congress allocated the funds and which handled the applications was Allergy and Infectious Diseases. Since many of the grant applications and some of the work sponsored by this institute dealt with phenomena that could not be accommodated under either allergy or infectious diseases, administrators of the National Institutes of Health suggested at Senate hearings that the name be changed to Institute of Microbiology. The idea was quickly squashed by a senator, who asked: "Who ever died of microbiology?" This is an example of a wrong and demagogic question. Had he asked, "How many lives have been saved by microbiology?" the answer would have been, "Many millions." Indeed, many millions died of plague and cholera because there was no microbiology before Pasteur, Koch, and others had discovered the causes and spread of infectious diseases.

For about two decades, this country has generously supported basic biomedical sciences. The money was administered by the National Institutes of Health and allocated by the peer review system. This program has propelled the United States into world leadership in the medical sciences. If you are suspicious of this pronouncement made by an American scientist, I invite you to go to any

71

state in Europe, to South America, or even to Russia, Asia, or Africa. Ask the scientists there where most of the advances are being made; ask them where they are sending their young, promising scientists for further training; ask them why journals published in Germany or Holland have most of the papers written in English, why the Japanese prefer to publish in bad English rather than in good Japanese, why even the Russians publish English summaries. We don't publish Russian summaries! Ask them why virtually all international symposia speeches are presented in English.

The answer you will receive is that American scientists have indeed dominated the field of medical sciences, and you will find that they are greatly respected for their know-how, reliability, and imagination.

The Right Questions and the Peer-Review System

These are what I believe are the right questions to ask about how to allocate funds. They are so simple that I hesitated when I wrote them down. All we need to know is who is doing good research in this country and how much money they need to do it. The answers to these questions are available and have been accumulated over the years by the most remarkable review system of the National Institutes of Health. I realize that this system has been criticized. Surely, mistakes have been made—because of human deficiencies—but I am certain that no better method than the peer review system is available to us. Many of the complaints have been made by persons who were disgruntled because they felt they had been unfairly treated. None of them has proposed a superior procedure that makes any sense to an intelligent person. I have served on these study sections as a member and chairman. Ten to eighteen scientists from all over the country with different viewpoints and backgrounds get together, often meeting each other for the first time. The debates and reviews of the applications often become quite

72

heated. The criticisms that this is a "club" operation among friends sounds very unfair to anyone who has sat in on some of these sessions. Moreover, the strictly enforced rules of rotation make certain that no permanent power structure can be established. I challenge anyone to come up with a design for a more impartial and competent decision-making body. There are scientists who will tell you that there are simpler and cheaper ways to do this. This is true. They tell you that these simpler and cheaper ways are just as effective. Are they? Unfortunately, I cannot prove that they are not because there is no objective test. We have to appeal to reason and common sense. For example, can an administrator deal objectively with hundreds of applications even when he is supplied with one or two letters for each applicant from experts who are not challenged in open debate? Outsiders do not realize that in the peer review system, as operated by the National Institutes of Health committees, each reviewer is on trial as well as the applicant. I have learned more about the capabilities and knowledge of my colleagues when I listened to their discussions of the work of other scientists than by reading their printed publications. I have accepted always with great reluctance appointments to such review committees, and I do not remember a single committee member who did not begrudge the time this task has taken away from his or her own research. My wife can testify that I returned virtually exhausted from these sessions that lasted twelve or more hours daily for two to three days, often over weekends.

I have spent much time talking about these committees because they have provided the information that we can use to arrive at the numbers we need. How much research should be supported? Let us not be concerned whether it is basic or applied but whether it is good or bad. If it is bad, it should not be supported. Supporting weak research is not only wasteful but harmful. Usually bad scientists attract bad students and give them bad training.

These second- and third-rate students who would not have otherwise made the grade are drawn into the practice of research and dilute its quality.

Recommendations

A more stable budget and decreased administrative costs. I propose that in determining the cost government should follow many of the recommendations outlined in the recent report of the President's Biomedical Research Panel. I propose that we have a permanent (but rotating) panel of such advisers who will provide the factual information and the budgetary items needed. They should propose a reasonable figure of the percentage of the Gross National Product that will support a healthy research program in this country. They can make recommendations on how to reduce the enormous cost of the administrative bureaucracy which cripples the finances of our universities. They can help to eliminate all the red tape about "effort input" and the "checking procedures" that force us to be dishonest because wrong questions are asked that cannot be answered honestly. They should help members of Congress to learn to trust scientists at least as much as they trust their colleagues in Congress. I have seen the headlines in the news about congressmen and scientists who misuse the trust of the public. But I am sure they are few in number, and they pay heavily with fear and with disgrace after their deceit is disclosed. I am sure that effective punishment after disclosure is better than constant checking. It is cheaper to trust than to check.

I believe that dishonesty among scientists is particularly rare. Non-scientists do not realize that dishonesty in a scientist is fatal. Dishonest scientists always choose important areas of research, otherwise no one would notice or reward them. Invariably, other scientists who want to develop these new discoveries try to repeat these experiments and fail. The glory of success that goes with faked experiments is very shortlived. You may wonder why—if this is true—these scientists publish work that brings them disgrace? Although I personally know a few such individuals, I do not know the answer to this question. I believe it is a psychological disease.

Whatever the reasons for fraud in science, the most important feature is that it is very rare. The simple reason for this is that crime does not pay in the field of science. Moreover, with the rapid growth of science, the survival time of fraudulent claims has precipitously dropped to a few years or even months and is followed by disgrace.

Better communication between basic and applied research scientists. Is there an excessive lag between basic discoveries and application? In the report of the President's Biomedical Research Panel, there is a scholarly review of the reasons for excessively long lags in the history of medicine. To be noted in particular is the impedance by regulating bodies such as the Food and Drug Administration. Can we improve the situation which presently is admittedly chaotic? In addition to the recommendation outlined in the report, I have a practical suggestion which might not only help to reduce excessive lags but allow us to tap the imagination of our scientists before some of their discoveries are published. I first proposed this idea to Dr. Philip Handler, president of the National Academy of Sciences, when we participated in an anniversary celebration of the Hebrew University in Jerusalem in 1965. This was long before the pressure for "applied science" was on. I had learned during this visit that the Israeli government employs physicists and chemists who visit experts in the basic sciences and discuss with them possible practical applications of their work. Many outstanding scientists do not have the talent of practical thinking. Men like Pasteur are the exception. The Israeli system obviously would not work in this country; we are too big. But there are simpler ways and I can think of two approaches that would cost very little.

I would appeal to scientists to present practical proposals that may or may not merit exploration by a commercial company. These proposals should be scrutinized by experts for scientific soundness. They can then go out for bids to institutions and commercial companies who can apply for government contracts to explore these projects. The originator should have a position as a consultant or collaborator, if he wishes. I realize there are legal problems that need consideration, and some form of simplified patent regulation may have to be devised.

Let me give an example of what I have in mind. After the recent outbreak of Legionnaires' disease in Philadelphia, a Public Health Service scientist suggested on television that a toxic "uncoupler" may have caused the disease. What is an uncoupler? I mentioned earlier that mitochondria are intracellular organelles which serve as the energy powerhouses of our cells. Uncouplers are compounds that short circuit the energy flow from these powerhouses. The energy is dissipated as heat, instead of being used to do work. Indeed, a high fever was the most common and unexplained symptom of the disease. If a toxic uncoupler was indeed responsible, how should the patients have been treated? Although I am supposed to be an expert on mitochondria and uncouplers, I was suddenly aware of the fact that I would have had no advice on treatment had I been asked. Cooling the patients and similar obvious symptomatic remedies are routinely used survival procedures. But how could we counteract these toxic drugs? I then realized that we were lacking basic information. Do patients or animals exposed to uncouplers die because they are internally burned to death, or do they die because their energy supply has reached a fatally low level? We don't know. If we did, we could design a therapy. If it is heat that kills, we can control the rate of combustion. But this would make the patient worse if the energy supply were at fault. In the latter case we could turn on an emergency energy generator in the cell which ferments glucose, is not mitochondrial, and not impeded by uncouplers. But this is a heat producing process which would make the patient worse if it is heat that kills.

I am not equipped to do such pharmacological experiments which will answer these questions, but I can think of them. Presently there is no mechanism to bridge the lag between ideas and execution. But I could write up a proposal on the mode of action of uncouplers which is scientifically sound and which could give us precise answers on how to treat a disease caused by a toxic uncoupler. I doubt that any company would be interested in the project as it stands because of the rarity of the disease. Perhaps they would do it if supported by a government contract. On the other hand, I have no difficulty in enlarging the project to make it more attractive. Uncouplers were used many years ago for the treatment of obesity—a logical approach—but unfortunately a dangerous one. As in the case of the MAO inhibitors, the drugs were used before the basic research was done. In fact, the chemical dinitrophenol was used by physicians for treatment of obesity before it was known that it was an uncoupler. A few fatalities reported in the late 1920s ended the eagerness of doctors to use it. But other drugs emerged, like thyroxin, with a similar history of danger and fatalities. Now the approach is discredited—rightly so—and yet it is so logical! Once more we should rethink the problem. I could propose the design of drugs that would act as uncouplers, yet we could reduce the danger of heat production by simultaneously depressing combustion. But first I need more basic information on the sequence of events in the cells that are triggered by an uncoupler before I can design a logical compound.

Training of young scientists. Important questions have been raised. Are we training too many scientists? Why should the government support their training? Are we discriminating against other students who go into arts or law?

I do not believe we are training too many scientists. I do

plead for special treatment of science and scientists when I ask for a stable budget and expenditure of funds needed for training. Such a plea needs a strong defense. I am convinced that the salvation of mankind will come through answers provided by science and an enlightened government that translates these answers into practical applications. For thousands of years governments have experimented with various social structures and have failed to devise one which does justice to all. They have failed to prevent discontent and the various forms of economic and behavioral diseases that have furnished the nourishment for the growth of a Hitler. They have failed to predict many of the consequences of urbanization and dehumanization by superstructures of civilization that have become modern towers of Babel with too many languages and lack of communication. Please don't misunderstand me. I do not advocate reversal of the time clock or a flight from reality, and I do not think we should blame technology or science for what we have chosen to do with the knowledge we acquired.

Members of Congress have two responsibilities. They are faced with many acute problems of the present, but they also have to look to the future. Some of my colleagues say that we are training too many scientists for the future, that there are not enough jobs available for them when they complete their degrees. The proposals I have made above will expand the opportunities for scientists and give them new dimensions. Could we train and inspire some of our students who are concerned with "relevance" to think in terms of the need to translate the information emerging from advances in basic sciences into projects of practical significance? I mentioned before that this gift of translation is rare, and I have searched for it among our students. A few years ago I offered a course at Cornell University under the somewhat pretentious title of Basic and Applied Science Correlation. From my contact with many students and postdoctoral fellows in the past ten years, I have come to the conclusion that probably less than one out of fifty scientists has real talent for bridging the gap between basic and applied science. But I believe it would be more effective to search for these talents and put them to work than to put pressure on basic scientists to focus their efforts on practical problems which require abilities that they do not have. If we can train a corps of scientists with a mission to reduce the lag between basic and applied science, we will be able to afford to let the basic scientists do what they do best and at the same time create a new profession of great value to society.

The proposal I have made for the allocation of funds for research, which is based on the principle of free enterprise and the survival of the fittest, can be applied to the problem of funds for the training of scientists. Let us make sure that all of our young women and men with a gift for science have an opportunity to do science. Let us keep the standard high and allow the survival of the fittest. The opportunities will grow with the scope and brilliance of their imagination.

There is much more to be said about the training of young scientists. Members of the President's Biomedical Research Panel have presented a thoughtful report and have made many practical suggestions. Most important is that we need a stable structure of funding and not one that flutters in the wind with the mood of Congress and the President. More than anything else, young scientists have become distrustful of government and uncertain of their purpose.

We need young scientists, and I believe that their efforts, particularly in the fields of mental health, behavior, and education will teach us about the causes of aggression and discontent. If we can learn to control these behavioral diseases, we can hope to build a peaceful and better world for its inhabitants.

Science—Risks and Benefits

It has become fashionable to blame "science" for many miseries in the world. People who previously opposed science because it was a luxury, now oppose it because they say it is an evil.

It is not difficult to document convincingly that science is neither a luxury nor an evil. Although we need to separate the concepts of science from science-derived technology, I do not say that science is pure and technology impure, for I believe that scientists have to share responsibility for the consequences of their discoveries. I shall discuss this at greater length, but here I wish to stress briefly the enormous benefits that society has reaped from advances of science and technology. I shall forego discussion of the acquired luxuries and conveniences of life, but are critics of science willing to return to the pollution of yesteryear, to the epidemics of cholera, plague, and typhoid fever, which have been wiped out by advances in sewage disposal, sanitation, and preventive medicine? Are they willing to accept the miseries of pain prior to anesthetics, of the crippling of polio, venereal disease, and other afflictions?

What has happened to our sense of fairness, to our common sense, and good judgment? Does society want perfect science? Let us admit that scientists have made mistakes, like everybody else, and they will make mistakes in the future. We cannot advance without doing so. To quote from Piet Hein's "Road to Wisdom":*

> *The road to wisdom?*
> *Well it's plain*
> *And simple to express . . .*
> *Err and err and err again*
> *But less and less and less. . . .*

Why is society becoming fearful of taking risks with future advances in science when our citizens are willing to take daily risks (and sometimes unnecessary risks) with past accomplishments of technology, such as cars, planes, and even snowmobiles?

A few thousand years ago, when our ancestors were in search of food and picked poisonous mushrooms, they died. Was that worth the risk? This is difficult to evaluate. They might have been hungry or even starving. We should be grateful to them because by trying and erring, by eating and dying, they conquered the peril of starvation, and their efforts led to the science of nutrition.

Today our society is beginning to turn against science because of fear of what the future might bring. Chemistry is feared because of pollution; physics, because of nuclear explosions; biology, because of genetic engineering; and medicine, because of toxic drugs.

Perhaps the greatest mistake we scientists have made is not to communicate better with the community in which we live and transmit a sense of optimism for a better world. I am an optimist. I believe that many diseases of society, including those created by advances in technology, can be cured by further advances in science. We need to convince our leaders in government that our defense of science is not self-seeking, that the benefits derived from science continue by far to outweigh the "evils."

* From *Grooks*, New York: Doubleday & Co., Inc.

We scientists realize that the mood of society is changing and that we have to account for our actions. Let us account. We can make a good case.

Recombinant DNA

Dr. I. I. Rabi, the well-known physicist, recently remarked that whereas scientists previously had to fight against biological warfare, they now have to fight warfare against biology. The activities of biologists appear to have become so menacing that Congress has considered legislation to restrict some areas of research in the field of recombinant DNA. The clouds of suspicion that have blurred the views of society about science in general because of all the "evils" that have followed advances in technology, have now settled over the field of biology.

The story starts with some early experiments with bacteria and viruses performed by Dr. Paul Berg, a brilliant scientist at Stanford University, in which he constructed a recombinant DNA molecule outside a living cell. Because some colleagues expressed concern about the consequences of such experiments, Berg became involved in organizing discussions about their potential dangers. At a scientific meeting in New Hampshire in 1973, the scientists decided to ask the National Academy of Sciences to examine the problem. Berg and Maxine Singer, a National Institutes of Health scientist, sent a letter to *Science* magazine voicing their concern. The Academy appointed a committee with Berg as chairman. In April 1974, Berg and members of his committee decided on an unprecedented move. They asked for a temporary ban on certain types of recombinant DNA experiments which they considered to be potentially dangerous, particularly those concerning tumor viruses and drug resistance of bacteria. Berg and his colleagues were aware of the fact that some of the molecular biologists who performed experiments with recombinant DNA had little experience in the handling of infectious agents. They felt that however remote the possibility of creating a new infectious agent, time was needed to sort things out, to make molecular biologists aware of the problem and to develop guidelines.

In February 1975 over one hundred scientists and a few lawyers met at Asilomar, California, to discuss the problems. Berg and the members of his committee, who presided over this conference, wrote a report in which they asked for the continuation of the ban and recommended to the National Institutes of Health specific and novel scientific and administrative procedures that would ensure safe conduct of recombinant DNA research. A committee appointed by the National Institutes of Health developed a set of guidelines for the protection of society against the potential risks.

The response of some scientists, of the press, and of the public to these demonstrations of restraint and responsibility was unexpectedly vehement. They reasoned that if the scientists themselves called a halt to their beloved research, there had to be the creation of a dangerous bacterium in the offing. The war on biology was on. Moreover, doubts were raised about the possible benefits of genetic engineering for the treatment of diseases, and it was even suggested that such manipulations may help to spread rather than cure cancer. Some scientists called for immediate cessation of experiments on recombinant DNA.

What are the dangers? There are two types of risks: the danger of the experiment itself and the danger of what we do with the acquired knowledge. Before discussing the dangers of recombinant DNA research, I should like to make a personal statement. First, I am not doing research on DNA. There is no conflict of interest. Second, I have had clinical experience. I was resident in an infectious diseases ward, and I have had laboratory experience with infectious agents. I taught microbiology in a medical school for nine years and each year between 1949 and 1952 I designed and conducted a three-month laboratory course on infectious animal viruses, including tumor viruses. I point out that among the most vocal opponents to recombinant DNA, there is not one I am aware of who has had experience with pathogenic bacteria or viruses.

Some opponents object to all work on recombinant DNA; others object specifically to the use of *E. coli*, a normal inhabitant of the human intestine. They recommend that other micro-organisms should be chosen. Since several reputable scientists have taken this stand, I would like to discuss it first. The reason for choosing a particular strain of *E. coli* (K12) was success. It is a very suitable organism for experimentation, and we know more about its genes than about those of any other bacteria. Moreover, the scientists working with the K12 strain have selected a mutant, EK2, that cannot survive in the intestinal tract of animals and humans even when administered in massive quantities.

In a workshop in Falmouth, Massachusetts, in June 1977, fifty invited participants met to discuss this problem. Many were experts in infectious diseases; some specialized in infections of the intestine, but most did not themselves work in the field of recombinant DNA. They came to a unanimous agreement that the enfeebled EK2 mutant of K12 could not be converted into a pathogen by a laboratory accident. One estimate was made that it would take about twenty years of fulltime research to accomplish this by *deliberate* attempts.

What is the basis of these remarkable statements? We know that the characteristics of infectious bacteria are very complex. No wonder that only a tiny minority of the micro-organisms that inhabit the earth are pathogenic. Bacteria causing infectious diseases must survive in the environment in order to spread. They must be able to enter the host by penetrating the skin or mucous membranes, or they must survive in the intestinal tract in

84

competition with other bacteria which are firmly established. If they can pass the barrier of the skin or the mucous membranes, they must still be able to spread in the host, resist its immunological defense mechanisms, and produce a toxin or cause some other form of damage to produce disease. There are many genes that control these properties required for pathogenicity, and the probability of conferring all those properties by accident is infinitesimally small. An experiment was made in which volunteers were fed a vaccine strain, a hybrid between K12 $E.$ $coli$ and a dysentery bacterium. Although about a billion bacteria were ingested, they were cleared in a few days, and no antibodies against dysentery were evoked. In contrast, the "wild-type" K12 $E.$ $coli$ strain was effective as a vaccine strain. The EK2 attenuated mutant strain of K12 did not even survive when fed to germ-free mice which are free of competing micro-organisms; EK2 is thus the only enteric bacterium known that cannot colonize in such animals.

Let us face the question of what would happen if a new infectious agent were actually created in the laboratory, a pathogen as virulent as the bacteria which cause plague or cholera. History may serve as witness. Even before we had effective antibiotics, the spread of plague and cholera was controlled by common sense. But common sense could not be used before we had an understanding of the disease, before we knew the micro-organisms that are responsible, or at least before we knew how the disease spreads to cause epidemics. Once these basic questions were answered, simple measures, such as control of food and water supplies as well as isolation procedures, effectively stopped the epidemics.

What about the first set of rules and guidelines set up by the National Institutes of Health to prevent the spreading of the hypothetical pathogen which might evolve from experiments with recombinant DNA? In view of the points made above, I believe these guidelines were over-protective and too restrictive. I know of important experiments that could not be done because of restrictions, e.g., putting a gene of a *Salmonella* bacterium back into the same bacterium. These restrictions may have hampered progress with the immensely important Ames test for mutagens which is carried out with *Salmonella* strains. I realize that the guidelines were in harmony with the mood of the time even though they were still insufficient for the extremists. Let us remember that we have already had about thirty years' experience with experiments now under restrictive guidelines and several years' experience with "dangerous" recombinant DNA research without the creation of a single dangerous pathogen. Let us also remember that recombinant DNA research has been performed by nature for millions of years. Why has a superbug with extraordinary survival value not evolved?

There is no question that guidelines and assessments are needed, but who should be responsible? Is it reasonable to expect Congress to understand the incredible complexities of DNA replication? Can any lay person without proper background give valuable counsel on safety measures concerning atomic energy? I believe in the value of including intelligent and dedicated non-scientists in such assessment and on regulation boards because there are often problems that may benefit from the non-partisan viewpoint. But let us not be blind to the fact that when it comes to technical details, the lay person has to rely on the integrity and good judgment of the scientists on the panel. It is therefore important that the composition of such a board be carefully balanced between specialists and those with a broad background. The necessity of avoiding conflict of interest among members unfortunately has been ignored too often. On the other hand, we have been impressed by the resistance shown by the Citizen's Experimentation Review Board of Cambridge, Massachusetts, made up entirely of non-scientists, to the political pressure by a mayor and by obsessed scientists. Nevertheless, let us note the narrow vote (5 to 4) of the City Council which replaced a politicized committee proposed by the mayor.

Guidelines, regulations, and legislation concerning research, even when generated by the wisest panel members, are likely to retard the progress of science. If absolutely necessary, this is the price we have to pay. But let us keep these restrictions to a minimum because they not only retard progress but are inevitably expensive. In the case of recombinant DNA, they are costly in time spent on administrative chores, construction of sometimes unnecessary containment facilities, failures of cooperation, misunderstandings or human error of judgment followed by punishments and embarrassments on all sides. They are costly because our most talented young scientists will be discouraged by the bureaucracy and the threat of punishments.

Hopes have been stirred that recombinant DNA might give us a treatment for genetic diseases, a cure for criminal tendencies, and mental disorders. I do not believe these hopes are realistic, even for decades. They are as unrealistic as the fears of meddling with the genetics of the human race, of the creation of Frankensteins* or Einsteins (some non-scientists are not sure which would be worse). The practical problems associated with such manipulations are monumental. Even the insertion of a gene into a plasmid for the production of insulin by *E. coli* is much more difficult than was first anticipated. But at least we have the technical knowledge to proceed, and in time these problems will be solved just as those of the landing on the moon were solved. The direct introduction of a gene into a chromosome of a human being is not solely a problem in technology. There are still too many

* In the original novel, M. W. Shelley's Frankenstein (a brilliant student) did not create a monster, but the reaction of society made him into one.

unknown factors. The unknown can only be unraveled by basic science and not by technology. A more indirect approach of gene therapy using an attenuated virus which carries the gene seems more practical, but is beset by other difficulties. How will the gene reach the proper target cell and not others in which it may be harmful? How will it cope with the immunological defenses of the host? How can we control the dosage problem so that we can avoid the overproduction of the missing hormone or the missing enzyme?

The hopes are much greater for advances in the laboratory. Hormones such as somatostatin and insulin have already been made in test tubes and other hormones will soon follow. Economically such productions are much more advantageous than chemical synthesis. The production of antibiotics will receive an enormous boost by genetic engineering. Costs of chemotherapeutic treatments should drop considerably. I can see a new future for antibodies against viruses and bacteria with the prospect of making proteins that are not antigenic.

Recombinant DNA research will undoubtedly give us new insight in the understanding of the chromosomes of plants, animals, and man. It will enable us to isolate single genes or groups of genes in high purity and large quantities. It will enable us to join together different DNA genes that will help us to understand phenomena of interaction and of differentiation, which are still mysteries. It should enable us to use these isolated genes for diagnostic tests and as probes for the elucidation of important problems such as gene expression.

I know that pessimistic views have been voiced about the possible impact of genetic engineering for agriculture. Still, I believe that this is a field for which genetics and recombinant DNA research holds great promise. It is regrettable how very few centers of excellence there are in the area of nitrogen fixation. If recombinant DNA work could increase the efficiency of nitrogen fixation in soybeans by only 20 percent it would help the food problem. If nitrogen-fixing bacteria could be adapted to plants that now depend exclusively on fertilizers, the impact on the economy and availability of food products would be enormous.

Commercial companies have taken up work that started in Israel on the use of micro-organisms to degrade oil (e.g., in oil spills). Genetic engineering could greatly increase the effectiveness of these bacteria and also insure control to prevent their spread into fuel reservoirs.

I believe that in the area of recombinant DNA work, the combined imagination of our scientific community falls short of the benefits we shall eventually derive. It is futile to make promises of the cure of cancer which we cannot fulfill because of our limited knowledge of this disease. But I am bold enough to predict that recombinant DNA work will have an effect on this disease indirectly. I also believe that micro-organisms and genetic engineering will be increasingly used to clean up our environment. Thus we shall find ways to combat the evils of science with science.

Pollution

There is no denying that chemistry has polluted our world with toxic compounds. In 1972 about six million tons of chlorinated hydrocarbons were manufactured and used in industry for dozens of different purposes. Many of these are mutagenic as well as carcinogenic and can be detected in drinking water and in human fat. Are the quantities large enough to cause cancer? We do not know and may not know the answer for another ten years if the pattern of the connection between smoking and lung cancer is taken as a model. Since 1960 the use of some of these compounds, such as ethylene dichloride and vinyl

89

chloride, have more than tripled and no one knows exactly what the effective carcinogenic dose is likely to be. It probably will vary with individuals.

But let us look at the other side of the coin. How much "pollution" has scientific technology eliminated? Nature is full of poisonous substances. Let me start with a small problem: peanuts. Many plants, particularly peanuts, may be infected with molds that produce aflatoxin and other toxic compounds. Aflatoxin has been studied extensively only since 1960 when it caused an outbreak of turkey X disease in England with about 100,000 turkey fatalities. This very toxic compound is one of the most potent carcinogens we know. It is already effective when present in the food of rats in ten to twenty parts per billion (about 0.2 micrograms per day). Other toxins formed by molds have been found in corn, wheat, rice, and soybeans throughout the world, particularly in Japan and Russia, and the diseases caused by them have been referred to by some scientists as "the neglected diseases." Do we have unlimited confidence in the tests that are performed and in the "allowable levels" of aflatoxin which are present in commercial products? No, we are not that stupid, but we are willing to take these risks. What about the most dangerous pollution of all? What about the infectious micro-organisms in the air, in water, and in food? Are the dreamers who want to do away with science and return to "nature" willing to accept pollution by cholera, plague, and typhoid fever? In the past hundred years, a brief span in the history of mankind, life expectancy has increased from thirty-six to seventy-two years. How did this come about? By discoveries of micro-organisms and their mode of spread, sanitation, preventive medicine, and the use of drugs. Are those people who are bemoaning the evils of science willing to accept a life expectancy of thirty-six years? Are they willing to return to the high infant mortality of the last century? Are the proponents of

90

natural food willing to drink unpasteurized milk and take the risk of catching typhoid fever or brucellosis? Or do they only want the benefits of science and technology without the risks?

When the polio vaccine was first developed, we had to take risks. In the early days of vaccine production, a mistake was made. The victims who acquired polio had to pay the price, and the company who made the bad vaccine still feels the blow of the crippling catastrophe. Can we hope that we can ever eliminate all human errors? Should we use such failings as an argument against new vaccines or new drugs? In the case of poliomyelitis, society decided that the benefits outweighed the risks and continued with vaccination.

What about the manmade pollution by industry, cars, and planes? Obviously we have to fight it, but we should fight it by making better cars and planes. This will not be possible unless we can convince the public that increased costs of these benefits are worthwhile. We have to fight pollution by making both the public and industry aware of the toxicity of chemicals that are produced and disposed of. When chemicals prove to be toxic, we must search for better ones that are not. This can be done and is being done.

Earlier I mentioned that Dr. Bruce Ames developed a sensitive test for the detection of mutagenic substances in a simple assay with bacteria. Hundreds of commercial companies have adopted his methods to test the chemicals which they produce and use. This is a very important step in the right direction, an achievement which was possible because an ingenious scientist had a brilliant idea. But he could not have developed this idea if he had not spent the first twenty years of his scientific career on problems of basic science allowing him to devise this test.

His discovery of a potent mutagen in female hair dyes was publicized in the daily press. It was bad publicity for the companies; it was bad for business. If a company can make a better dye which is not mutagenic, it will conquer the market. Education can be a powerful weapon.

Drugs and Food

Fear has played a dominant role in the development of new drugs for the treatment of diseases. The public, inflamed by the press, has forced administrators into a position of timidity and caution that is detrimental to the progress of medicine.

Could Pasteur have done his bold experiments today? Could we have progressed today with chemotherapeutic agents as rapidly as we did thirty to forty years ago? Would prontosil, the forerunner of sulfonamides, be passed today as an acceptable drug? I think the answer to most of these questions will have to be no.

Our society has become a society of the fearful; we are afraid of the food we eat, of the water we drink, and of the drugs we take. The clamor to return to natural foods is no more reasonable than the fear of new drugs. Additives are called evil. People forget the enormous benefits, the food saved, and the prevention of bacterial growth. Natural plants, non-edible as well as edible, are full of chemicals that are toxic and carcinogenic and provide us with no benefits. I mentioned earlier the mold which produces aflatoxin. It is as natural as the mold which produces penicillin. Should we put a ban on molds because they produce toxins, or should we keep studying them and select those which are beneficial and discard those which are harmful? The answer to this question is obvious to everyone, but we are not willing to apply the same answer to recombinant DNA or to the development of drugs. How did our ancestors find out which food is edible and which is not? As I said before, by eating and dying. In our search for new drugs, we have made it easier for ourselves. We have let mice and guinea pigs die for us first.

This procedure has saved many lives, but in the last analysis man is the only model for man, and we have to take the risks of the final test.

How do we stand in the world today in the testing of new drugs? Dr. Donald Kennedy, the head of the Food and Drug Administration, has written an article on this subject entitled "A Calm Look at 'Drug Lag,' " which appeared in the *Journal of the American Medical Association* (239: 423, 1978). Although I have great admiration for Dr. Kennedy, I do not share his view on the "drug lag." I have no shares in the pharmaceutical industry, am not a consultant, and am willing to respond to his challenge. Yes, I am "willing to step forward and ask the Food and Drug Administration to be less vigilant about safety," particularly when the "vigilance" includes unnecessary hurdles of red tape. As we can read from Dr. Kennedy's statistics, the United States introduced a total of fifteen new drugs in 1976, compared to twenty-one in England, twenty-three in Italy, thirty-eight in France, and thirty-nine in Germany. I do not propose that we reduce our rules to those operative in Germany, which are probably too lenient, but I do believe that our pharmaceutical companies could offer more than fifteen promising new drugs in one year. Numbers of this kind mean little, but it is a fact that the approval of new drugs has declined in the United States since 1962. The time and cost to approve a drug have increased excessively, beyond the expected increase from inflation. It costs millions of dollars to put a new drug on the market, and small companies cannot afford this financial burden. Even large companies hesitate to develop drugs for minor diseases because they cannot count on a financial gain. Drugs are often approved first in other countries before they are approved here. Sometimes, as in the case of an effective antiepileptic drug, there was a delay of many years before approval was given. These are the facts properly cited by Dr. Kennedy, yet he says they make a "superficial case." I am looking forward to the fulfillment of his promise to change some of the policies that unnecessarily delay the approval of new drugs and to the "greater flexibility in the clinical testing and marketing of new drug products."

The Basic Question

The basic question society has to answer is whether we should stop studying certain aspects of nature (indeed, DNA recombinants do occur in nature) because of a possible danger or because of fear of what we will do with this knowledge. This is a question we have to answer for all explorations into biology, chemistry, atomic physics, and other sciences. It is a difficult question to answer because we cannot foresee what the future will bring. We cannot apply science to answer this question; we have to approach it with judgment of values; we have to approach it with faith. And I have faith—and I see no alternative— that the solutions to most problems—physical, economical, and psychological; problems of crime and war; hate, discrimination, and oppression—are dependent on our understanding of their causes. History tells us that for over thousands of years we have made little progress in solving basic social problems. On the other hand, look at the extraordinary progress made in medicine during the past hundred years because of advances in biology. In this short span of history, we have achieved a doubling in our life expectancy. From this viewpoint, the perishing of a few scientists working with dangerous viruses or rickettsiae, the death of some patients hypersensitive to a chemotherapeutic agent, or even the release of some drugs that proved to be unsafe become tragic incidents that we must accept as the price we have to pay. I believe that the solutions to most problems besetting the human race will more likely come from scientific progress in the fields of biology, psychology, economics, and social behavior than from political revolutions.

94

As astronomy has emerged from astrology, chemistry from alchemy, we are now witnessing the emergence of the new science of sociobiology which is built on important studies of social behavior in animals. Even bacteria now serve as models! They swim toward attractants and away from repellants, patterns of behavior that may shed light on sensory responses, memory, and other important functions of our central nervous system. One day we may even have a science of psychobiology or molecular psychology.

This is where my viewpoint differs from that of Senator Kennedy. The chances are slim that we shall learn to cure many diseases by accidental discoveries, that we shall solve economic problems by an arbitrary scheme, or that we shall learn to solve the problems of crime and war without understanding their causes.

Epilogue

It is the business of science to step forward with dangerous ideas and to prevent and cure the evils that emerge from the technology that is derived from these ideas.

I realize that the views expressed in some of these letters may sound extreme and could be easily misunderstood. I do not propose that we wait for science to attack current diseases of society. I do not advocate this any more than I would recommend no treatment for medical diseases that we do not fully understand. As illnesses of the body have been treated by great physicians, diseases of society have been successfully attacked by great men in government. As we face the threatening problem of overpopulation today, we cannot wait for basic knowledge in the development of the egg. We must deal with criminals before we understand criminality. We cannot postpone planning for better health care until we understand all diseases.

The scientific approach does not relieve us of the responsibility to make value judgments that require us to attack social problems of our day. For this we need an enlightened government, members of Congress with vision and courage. I hope they will call more and more on scientists for advice, and I believe it is the duty of scientists to be available when needed.

What I have been talking about in these letters is not out of concern for today as opposed to tomorrow. I am thinking about the future, a future which, we hope, will provide a better world for our grandchildren and their grandchildren. To make such a world, not just in terms of a longer and healthier life, but in terms of peace and better mental health, we need to know more—more about aggression and ambition, more about desires to rule and willingness to be ruled.

When we have learned more about ourselves, we shall need—desperately need—enlightened leaders in government to put this knowledge to good use. Scientists must learn to communicate their knowledge more effectively to the representatives of society and to the public. We must convince them that the investment in science is in the interest of all and just a fraction of what we spend for the preparation (or preparedness) for war or for luxuries.

It is difficult to be convincing when the risks are highly visible to the public while the benefits are taken for granted. It seems easier to look back at a past free of pollution than at a past with plague, cholera, and polio.

We scientists have failed to project a balanced picture of science and scientists to the public, and if television programs do not change the image of scientists, we will fare no better with the next generation.

Scientists need to know more about themselves and their role in society.

I want to close with a quotation from an eloquent article by Dr. Lewis Thomas, which appeared in a recent issue of *Daedalus* (Summer 1977).

> Only two centuries ago we could explain everything about everything, out of pure reason, and now most of that elaborate and harmonious structure has come apart before our eyes. We are *dumb*.
>
> This is, in a certain sense, a health problem after all. For as long as we are bewildered by the mystery of ourselves, and confused by the strangeness of our uncomfortable connection to all the rest of life, and dumbfounded by the inscrutability of our own minds, we cannot be said to be healthy animals in today's world.
>
> We need to know more. To come to realize this is what this seemingly inconclusive century has been all about. We have discovered how to ask important questions, and now we really do need, as an urgent

matter, for the sake of our civilization, to obtain some answers. We now know that we cannot do this any longer by searching our minds, for there is not enough there to search, nor can we find the truth by guessing at it or by making up stories for ourselves. We cannot stop where we are, stuck with today's level of understanding, nor can we go back. I do not see that we have a real choice in this, for I can see only the one way ahead. We need science, more and better science, not for its technology, not for leisure, not even for health or longevity, but for the hope of wisdom which our kind of culture must acquire for its survival.

Index

101

LIBRARY OF CONGRESS CATALOGING IN PUBLICATION DATA

Racker, Efraim, 1913-
    Science and the cure of diseases.

    Includes index.
    1. Medical research.   2. Medical research—
United States.   I. Title. [DNLM: 1. Medicine—
United States.   2. Research—United States.
3. Science—United States. W20.5 R122a]
R850.R32   610′.7′2073   79-84012
ISBN 0-691-08243-X
ISBN 0-691-02363-8 pbk.